ASTEROIDS
in HUMAN DESIGN

Awakening the
Feminine
Archetype

KIM GOULD

Edited by Debby Levering
Cover design by Kristina Edstrom

An Imprint for GracePoint Publishing (www.GracePointPublishing.com)

GracePoint Matrix, LLC
624 S. Cascade Ave, Suite 201
Colorado Springs, CO 80903
www.GracePointMatrix.com
Email: Admin@GracePointMatrix.com

SAN # 991-6032

A Library of Congress Control Number has been requested and is pending.

ISBN: (Paperback) 978-1-955272-47-6
eISBN: 978-1-955272-48-3

Books may be purchased for educational, business, or sales promotional use.
For bulk order requests and price schedule contact:
Orders@GracePointPublishing.com

Contents

Growing up, we say, as though we were trees, as though altitude was all that there was to be gained, but so much of the process is growing whole as the fragments are gathered, the patterns found.

—Rebecca Solnit

It has taken me longer than expected to write this book, as I thrashed around in the shadowlands of a post-patriarchal world trying to find a perspective that would stand still long enough for me to write about it.

I have lived for six decades as a middle-class white woman unaware of her own privilege. As our modern world shakes and shudders I find myself searching for what comes next, while also realising how deeply and unconsciously we continue to be held within the world view of old white men.

I have the same struggle with the Human Design System. I can't count how many times I have described it as the synthesis of four ancient esoteric wisdoms. And yet each of these wisdom systems is built firmly on patriarchal principles because they have all arisen in the last 5,000 years. These principles permeate what I believe is a profound tool of consciousness given as a gift to guide us to the next level. Using the asteroids as part of my Human Design toolkit has helped me to find a way past some of that patriarchal filtering.

We are resting on an ocean of millions of years of peaceful human existence, buffered by a relatively short 5000-year tsunami of violence that threatens to snuff humanity out. Writing this book, I have been immersed in the mythology of pre-patriarchal humanity. This knowledge has left me hungry for a different way of being.

At first I was looking through a feminist lens, but as I dropped more deeply I began to see all the First Nation peoples, the colonised, the children who are afraid of losing their planet, the low-paid workers who can't afford housing, people who mine rare metals for my smartphone. It goes on and on. We have somehow come to organise ourselves around cultural stories that were mostly created during the times of Ancient Greece and Rome and are instantly recognisable to us. Zeus with his thunderbolt; Hera with her jealousy. But these endemic myths threaten to enslave and ultimately destroy us. They damage our very connection to reality, to Gaia, and to each other.

I believe we are in a process of recovering the repressed energy of those millions of years of human evolution, when we co-created the human ways of living and loving, caring and connecting. The message from some quarters is that we must be strong, selfish, even cruel, if we want to survive. The evidence is all the other way. It's wild love—for ourselves, for each other, for nature—that is what will move us through these times.

There is something seeking to erupt in the human spirit. It is protective, natural, primal, and fiercely alive. It is the raw potent energy of the long-suppressed feminine, and I dedicate this book to her.

———•◆•———

I remember as a young woman, finishing school and heading out into the world looking for my first job, I was shocked to find that there were things I wasn't supposed to be able to do. I discovered there were lots of jobs girls weren't suited for. Who knew I wasn't supposed to be independent and confident? Not me! I was confused when people expected me to put limits on myself as a young woman. My parents had failed to teach me those ways of diminishing myself!

Understanding the asteroids in my Human Design has helped me find my way back to that girl, who didn't know she was supposed to comply and make herself small in the world. As a young Artemis, I loved to wander in the forest with my dog. As a young Pallas, I discovered a love for law and justice. As a young Aphrodite, I flourished when I found love.

One winter morning, when I was a young twenty-nine-year-old lawyer, I collapsed in court. I'd finished my case and was about to head back to my office when everything went blurry and wobbly. I sat down for a while till I could walk, got a taxi home, and went to bed. I didn't get up for three years. I was diagnosed with ME/CFS (myalgic encephalomyelitis/chronic fatigue syndrome); it was as if I was in a coma. My husband had to wake me up to feed me or I would have slept twenty-four hours a day. At the time the asteroid Titania (queen of the fairies) was very active in my Design. It truly felt as if I had been taken to another world.

Overnight I had gone from commuting into the city each day and the busyness of court, meetings, and phone calls, to a very different reality. I realised this was an initiation. I didn't have the energy to cry, but it was a tough adjustment. I had to find a way to let go of everything I had thought my life would be. I had to go within and discover a new version of myself. As I was able to do a bit more I learned to meditate and began journaling, finding wormholes into other realities through my own felt sense of self. This was a new kind of adventure, and I slowly learned how to flourish there.

I was nourished by this newfound sense of truth, but it also felt isolating. I didn't fit anymore. I had to leave my people behind in order to claim it. I had to go alone. I had to shed the layers that no longer felt real. I got divorced. My children spent two years with their father. I moved to a wilderness area, the World Heritage-listed

Border Ranges in Northern New South Wales. I longed for a direction aligned with my renewed sense of purpose. I recognise this stage in so many people I work with now, the impatience to find that "thing" that reveals our new life. But I just had to be patient as I gradually became a different kind of person.

In 2003 I saw the Human Design mandala in a magazine and knew this was what I'd been waiting for. I spent the winter with my Human Design course materials spread out on the floor of my living room in front of a roaring fire, unpacking this new knowledge. I know many of you will understand when I say I couldn't get enough of Human Design. I didn't want to leave the house; I was almost inhaling it. But after a few months, I became frustrated. I wanted it to shimmer with life, but it seemed fixed, set in its ways. I wanted it to open itself up to me, but it would only let me meet it on the surface. There were astonishing new insights, but also places where I felt it trapped me in an outworn way of viewing the world.

I needed to delve more deeply. I went exploring.

I started playing around with the planets in my Design, as a way of getting more personal. After all, graceful Venus is going to give me a very different experience of a gate than feisty Mars! One day I began to wonder about my ascendant. The ascendant is considered of major importance in astrology. Could it help me crack through to a new level of the understanding of my Design? I figured out my ascendant is in the gate that bridges my split definition. It brought my two parts together! It connected my four motors with my undefined Throat! I wasn't sure what this meant, but I could feel my feet were walking me towards something important.

After this auspicious break from Human Design orthodoxy, I ventured into an exploration of the amazing world of asteroids. I began by manually calculating the position of a whole long list of asteroids in my Design—archetypes like Persephone and Psyche, Pandora and Pallas. It took me a while, but I wasn't daunted, I had overcome these kinds of challenges before. In my first year of learning Human Design, I had an Apple computer and the only software available at the time ran on Windows, so I had to calculate all my charts manually and draw them by hand. That experience became invaluable as I headed off into these new unexplored dimensions of Human Design.

Grasping my precious list of about fifty asteroids, I set out to make some sense of all this new information. I pulled out my old copy of Demetra George's book, *Asteroid Goddesses*, and read about the importance of these feminine archetypes at a time when women were stepping into a different kind of leadership role in society.

Early in my asteroid quest I simply expected to find more information than I could get from a "normal" reading of a Human Design chart, perhaps deeper and more precise. What I didn't expect was the personal undoing they initiated in me. The gatekeeper, if you like, was Pallas. A large asteroid transiting between Mars and Jupiter, Pallas is a powerful cultural archetype. It was her story of being taken from her mother that broke me wide open.

The story goes that Pallas was born, fully formed and wearing armour, from the head of her father Zeus. But there's more (as you will learn, there is always more to these stories). Zeus had been warned that one of his children would be his undoing. When Metis became pregnant with his child, Zeus went on the offensive and swallowed her in an attempt to take control. Metis, knowing her daughter would have to go alone into the harsh world of men, put protective armour on her unborn daughter.

The hero version of the story is that the wise and skilled Pallas burst fully grown and covered in armour from the head of her father, as if ready to go into battle. I was reminded of my lawyer days, dressed in a suit, heading into court, tough and uncompromising. The young Pallas had no tender time to grow into herself, she was born knowing the rules and how to use them to her advantage. Ripped from her matriarchal roots, she mastered her new world, as many of us do.

Learning about Pallas in my Human Design took me on a personal journey of deep grief as I realised the yearning within me for something which at first seemed undefinable, but later I came to experience as the Great Mother. It was less a return than a recognition of an empty place where I intuitively felt I should have found safety and nourishment. As I allowed myself to rest into this strange and uncomfortable new space, I began to see hidden parts of myself and their confusion of not having the opportunity to naturally learn and grow. I held an internalised expectation that I would always show up fully formed. I could feel the heaviness of the armour that covered my sensitivity as I ventured into the world each day. I could feel the constant sense of failure that came from not being a man. It didn't matter how successful I was as a lawyer, I was always judged as not being good enough. So I worked harder and harder, not realising that I was mostly judged by my gender rather than my achievements.

Looking back on this time, armed with my newfound sense of the true history of the young lawyer Pallas, I could feel not just the loss of MOTHER, but the loss of the ground she had stood on. I didn't know how to stand on the earth as an adult woman and claim space, and my mother hadn't taught me because she didn't know either.

As my gateway to a new view of Human Design, Pallas taught me, over the years through my work with my clients, that women had adapted incredibly well to standing on patriarchal ground. We have expertly learned the ways of men and have grown beyond the restrictions. We are ready to throw off the armour, to stop using our gifts in the service of a culture of dominance and denial. It's time to move beyond the polarity of feminine (matriarchy) and masculine (patriarchy) and create a new kind of culture.

This book is not just about women reclaiming their place. Men are just as limited by a culture based on the domination of nature, the feminine, the gentle. Zeus, generally considered the *numero uno* in Greek mythology, seems to have spent his time on various forms of conquest, seduction, rape, and war. If this is the archetype for masculinity, how does any man find peace?

At this crossroads of human evolution, we have to find a way forward. The seeds for that pathway lie within our original human nature, and we can find clues to that original nature in the asteroids.

Feminine DNA

In the most common retelling of the story of Pallas, she is born from Zeus alone and has no mother. I can only imagine the pain of Metis, knowing her child will be lost to her, no one will even remember her daughter had a mother.

One morning, sitting in a local cafe contemplating the abduction of the daughter and the empty void where the mother should be, I was overcome with an idea. I asked the waitress to bring me a napkin and pen. I could hardly contain myself, and when she dropped them on my table, I grabbed the pen and wrote furiously in capital letters-WHAT THEY STOLE FROM US. It was an impassioned plea for the recovery of women's creative power, stolen away by the narrative of the patriarchal church and Abrahamic religions as they sought to make men the sole creators, to put men alone on the throne.

The story of Pallas revealed for me a simple hidden truth: We had lost both our mothers and our daughters. We had lost even the right to believe in a mother who was anything other than a passive and empty vessel to be shaped by the male "head." In her book, *Motherpeace: A Way to the Goddess Through Myth, Art and Tarot*, Vicky Noble describes the matriarchy as a regenerative culture, where we recognise the power of crisis to transform. The medial crone Hekate stands at the crossroads when we face an emergency, pointing the way for us to recover the fecund creative potential of the daughter. Without her we are lost in constant crisis, fearing change simply because the masculine alone cannot create new life.

Lest you think this loss of the feminine and indigenous aspects of culture just somehow naturally disappeared, Barbara Walker explains in her book *The Secrets of The Tarot: Origins, History and Symbolism,* that the Inquisition undertook a reign of terror in Europe for nearly 500 years to force pagan beliefs and feminine wisdom underground. She quotes Martin of Braga condemning women for "decorating tables, wearing laurels, taking omens, putting fruit and wine on the log on the hearth and bread in the well, what are these but worship of the devil? Calling upon Minerva when you spin, overseeing the day of Venus at weddings and whenever you go out on the public highway, what is that but worship of the devil?"[1]

---

[1] ISBN: 978-0062509277
Barbara G. Walker, *The Secrets of The Tarot: Origins, History and Symbolism,* Part I: The Sacred Tarot, 2017, US Games Systems, Connecticut, Kindle Version

This is a defining moment for humanity. We do not have time to continue to make do with the stories a few men tell us about our experiences. Women and men can no longer distort themselves in an attempt to fit within these destructive narratives about who we are and how we should create our lives.

The asteroid layer in your Human Design is a historical map for stripping away the outgrown limits of patriarchal narratives and recovering our original selves.

What Stories Does Our Human Design Tell Us?

The conventional thirteen activations in your standard Human Design chart contain only two feminine archetypes—Venus and the Moon. It is impossible that these two objects can carry the rich spread of the experience of being a woman in the 21st century. Where is our feminine DNA? We can find it in the asteroid layer of our Human Design.

Here we find the Great Mother (Ceres), the sovereignty of the Queen-Goddess (Isis, Aphrodite), the wild untamed Daughter (Artemis), and the power of the Sorceress (Circe, Medea, Hekate). Here we find the aspects of the feminine lost and now returning. It's time we began to live through the lens of our own eyes, and not through those of a philosopher-poet from Ancient Greece.

You may have seen people referring to some of the traditionally masculine planets as holding an emerging feminine energy. Saturn, Neptune, and Pluto are often viewed as having feminine aspects. We can certainly learn from this expanded perspective of the planetary archetypes, but their archetypal roots remain in the masculine sphere. Historically, the mythological Saturn was definitely not a female archetype, in fact, he is a fundamental part of the nature of patriarchy woven so deeply into our modern culture we often don't see him at all.

It's not just the feminine aspects of self we find in this more subtle layer of our Human Design. Asteroids like Eros, Apollo, and Orpheus fill in the gaps between masculine planets. Apollo is often described as a leader, but everything he had came from his father Zeus, and he suppressed his true nature to keep that patronage.

Apollo speaks powerfully to men's experience of becoming adults. The Ceres activation is just as relevant to men when it comes to understanding what nourishes us and how we have learned to value ourselves in relationship.

If our Human Design is a map of our soul DNA, a guide to the synergy between our unique personal energy field and universal consciousness, why not include such a rich treasury of information?

Asteroids in Human Design

In the original Human Design revelation in 1987, Ra Uru Hu was given a structure (the Human Design chart) created from a synthesis of four ancient systems. This was the year of the first Harmonic Convergence, said to usher in a new epoch of peace and harmony. Human Design reflects this global focus, bringing together the spiritual wisdom of humanity, providing a self-teaching tool for higher consciousness, connection, and purpose. This gift to human consciousness is intended to speed up our evolution, to support us to access a new integrated consciousness without having to spend decades and lifetimes being initiated into deep spiritual traditions. This would be impossible within our current evolutionary time frames.

Those four ancient systems give us the gates (Chinese I Ching), the channels and Centers (Jewish Kabbalah and Hindu Chakras), and the personal activations (Western astrology). It is the *synthesis* of a millennium of human esoteric wisdom embedded in these four wisdom traditions that creates the power and precision of the Human Design System.

The word synthesis is important here. The meaning of any part of your Human Design chart comes from taking all four components and blending them to find an entirely new meaning. I describe Human Design as emergent, which means once you put the parts together you get much more than you started with. We can look at a hexagram in the I Ching, say Gate 41, and it will tell us it is called Decrease and that it is based on an image and a story. Each hexagram is made of two trigrams that hold elements. But Gate 41 is also in the Root Center, so we have to draw in the meaning of the Root Chakra and the energy center Malkuth in the Kabbalah. There are structural relationships as well. What does it mean that Gate 41 meets up with Gate 30 for example? How does that add to the meaning of the original hexagram? And then we have the streams and circuits of Human Design to add even more meaning.

You will notice as you read about the asteroids that their discovery charts give astonishingly accurate information about their meaning in your Design. It's the same with, for example, the gates. There is a confluence of meaning when we put all these elements together.

This is not to say we should ignore the original meaning of the hexagram, or chakra and Sephiroth.

Scholars have been exploring all four of these esoteric sciences for thousands of years, and we impoverish our understanding of Human Design if we disregard that work and the accumulated wisdom that has arisen out of it. To neglect, for example, the thousands of years of exploration of the I Ching hexagrams is to deny ourselves a depth of meaning about the Human Design gates that just can't be replaced by using a single keyword or phrase.

If we take an attitude of learning the keywords for each part of our Design, we miss the very heart of this soul technology. Learning is only our doorway into the knowledge. We must experiment with it, allow its emergent nature to support our growth. This is a two-way process, a lifetime's dedication to creating a different kind of humanity. As you learn more about your own Human Design through experimentation, the whole quantum field of Human Design becomes richer and more conscious.

It's the same with the asteroids. People come to me looking for a set meaning. Aphrodite means love and beauty. Arachne means pride and skill. These simple meanings may be useful sometimes. But the meaning of asteroids is rooted in mythology, and the mythology is rich, diverse, and full of personal meaning.

You and I will have quite different experiences of the meaning of our Aphrodite activation. Her mythology speaks to us of many things—love, sex, pleasure, art, fashion, beauty, the ocean, and animals. We can also explore her influence through the lens of her relationships with Mars, Hephaestus, Eros, and Psyche. We will find what is important for us only by experience and exploration of the full archetype. If you have Aphrodite in the Sacral Center, you may find her expression through sexual pleasure. Aphrodite in the Heart Center may find herself looking for a rich husband who can shower her with expensive gifts. In the Spleen Center Aphrodite may express through style, fashion, and beauty. If your Aphrodite shares a gate with Eros, you will experience her energy very differently than if she shared a gate with Psyche.

Neither can we expect our personal experience of Aphrodite to remain static over time. There are many ways your Aphrodite can shift and change. If Pluto transits over your Aphrodite, you must go deeper into her world, allowing it to permanently transform you. Pleasure, playfulness, beauty, these qualities will burst through your daily life, demanding a new and more important place. Also, the nature of the archetype will change for us as we age. We will experience Aphrodite differently when we are seventeen to how she turns up for us when we are forty.

Exploring the asteroids in Human Design demands that we have a deeper understanding of the structural components of our Design, and also the potential emergence of new ways of experiencing them. You have Aphrodite in Gate 40 in your Design and you may use the keyword *Deliverance*. What if you have Aphrodite in that gate? How do you even begin to make sense of these two completely separate ideas—Aphrodite and Deliverance? How do you connect with your experience of Aphrodite in Gate 40?

I'll be taking you through the process of understanding your asteroid activations, and along the way your understanding of the nature of Human Design may shift.

Asteroids Change the Way We Work with our Human Design

Should we use asteroids in Human Design? Surely we should stick to that one single chart with its thirteen planets that Ra was given in the revelation, and anything else is not really truly Human Design.

Well, actually, no.

In *The White Book*, also known as *From the Book of Letters*, written in 1995, Ra Uru Hu, says,

> *According to the Voice, no other objects in the heavens, other than the ones listed, can activate gates. This is not to say they don't have an effect. They will filter the neutrino stream like everything else, but they do not activate a gate. They will add their quality, provide insight.*

So the original revelation didn't expect that Human Design would be forever limited to that single 13-planet chart.

But what does it mean, to say they "add their quality and provide insight"?

———•◆•———

When we look at the asteroids, we have to think differently about what Human Design is, and how we interpret and apply the information.

There are two ways that using asteroids can change the way we work with Human Design.

Firstly, they shift our perception of how energy flows through our chart. There are very powerful archetypes sitting in gates in your chart that you would usually consider "empty" because they are white or undefined.

Secondly, when we discover the most significant asteroids in our Design, we find new perspectives on our life events and learn to tell our own story in new and different ways.

Asteroids Change the Way We Think About Energy Moving Around the Chart

Let's look at Taylor Swift. She has an undefined Solar Plexus Center, with her Design Sun in Gate 36. Her Persephone is in Gate 35. If Persephone gave definition, Swift would go from being a Projector with an undefined Solar Plexus Center to an Emotional Manifestor with the 35/36 channel defined.

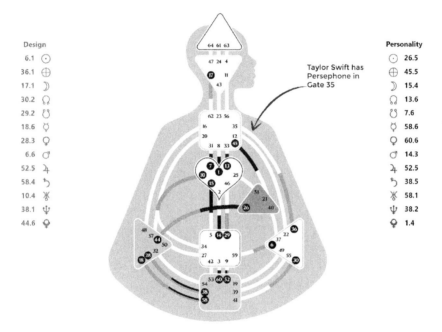

This becomes tricky in terms of Type. Is she really a Projector, or is she actually an Emotional Manifestor? Or perhaps she's a hybrid?

Most people don't realise that Type was not part of the original Human Design revelation, but a tool that Ra Uru Hu devised later to help people navigate the complexity that is Human Design. It's incredibly useful as a learning tool, but as soon as we step away from the limitations of that one single chart, our current fixed ways of thinking about Type quickly become problematic.

If I was working with Taylor Swift, I would definitely suggest that she experiment with waiting for emotional clarity. I wouldn't say that for every asteroid, some just don't carry that much influence in your life. The more significant an asteroid is, the more likely it will affect the way your Design operates on a structural level. Persephone is a powerful cultural archetype, signifying a deep wisdom about how to move through difficult times and come out on top.

There are two important things to understand here. Firstly, Swift has an undefined Solar Plexus Center so she will *always* experience her daily life as a person with an undefined Solar Plexus Center. Which means she feels the emotions of others in an amplified way.

But at an archetypal level, she is, perhaps literally, a channel for the voice of Persephone, for how women can connect with something powerful on an emotional level, find their unique perspective that gives their experiences, even the painful ones, real meaning and purpose. She is always acting on, and expressing her emotions through, the filter of that Persephone energy. So there is always potential emotional turbulence that comes from her Persephone expression.

There's even more to this story.

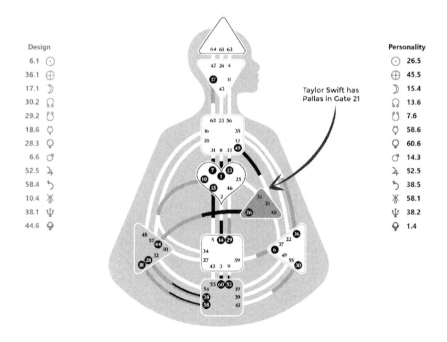

Design		Personality	
6.1	☉	26.5	☉
36.1	⊕	45.5	⊕
17.1	☽	15.4	☽
30.2	☊	13.6	☊
29.2	☋	7.6	☋
18.6	☿	58.6	☿
28.3	♀	60.6	♀
6.6	♂	14.3	♂
52.5	♃	52.5	♃
58.4	♄	38.5	♄
10.4	♅	58.1	♅
38.1	♆	38.2	♆
44.6	♇	1.4	♇

Taylor Swift has Pallas in Gate 21

Swift also has Pallas (creative brilliance, strategy) in Gate 21, bringing her defined Root and Spleen to the Throat via the Heart Center. If we take her Pallas activation into account, Swift becomes an Ego Manifestor.

So not only do her songs use the words of the Queen of the Underworld, Persephone, but Pallas gives her a brilliant strategic mind for business.

What if Swift came to you as an eleven-year-old saying she had persuaded her parents to move her whole family to Nashville because she knew that was the right thing for her career? Or if she was fourteen years old and described her decision to leave her RCA contract because they didn't believe enough in her? This was a young woman who knew how to successfully initiate action independently—the definition of a Manifestor.

If Swift was just working from her standard chart, she would be told she is a Splenic Projector and should wait for the invitation. But she has been initiating her whole life, and it seems to be working out pretty well for her. If she came to you to talk about her Human Design, how would you explain that to her?

It's not just about moving away from the fixed idea about Type. It's important for Swift to understand that she will have much less turbulence in her life if she pays attention to her emotions, and that she can trust her willpower to guide her business strategy.

Underneath all of that though, is the fundamental need for her to ensure she uses the gifts of her Projector channels and doesn't fall into the trap of letting her emotions or her will create situations where she has to work like a Generator.

Swift is such a great example of how the asteroids can completely shift our understanding of the structure of the chart, and of Type and Awareness. She is exactly the kind of Projector who comes to me to ask why her strategy isn't working for her!

Asteroids Change the Structure of Our Human Design

But if the asteroids Persephone and Pallas don't bring definition, why even bother to look at where they are in our Design?

Because, as the Voice said, these asteroids filter the neutrino stream in the same way all the planets do and have an effect on how we experience the energy of a gate. They also have a subtle influence on the Centers. For Swift, Gate 36 holds potential emotional responses to new experiences. In order to understand those experiences, and make progress, Swift needs to access the insight Persephone brings.

Persephone in Gate 35 won't BRING those new experiences, but when they arrive, it is Persephone who helps Taylor Swift figure out what to do with them, how to make meaning of them, how to express them, and how to learn from them. We can still see her Gate 35 as a white gate, and she will still be, to a greater or lesser extent, conditioned by people who have it defined, but that Persephone energy is always influencing how it operates in her life.

The asteroids have a subtle but important effect on how the energy flows through our Design, and that may upset our understanding of Type and Awareness and require us to experiment at a more subtle level.

———■·♦·■———

As well as subtly shifting the energy flow through the structure of our Design, the asteroids also help us bring to life stories and experiences that can't be explained by the thirteen standard planetary activations from that single Human Design chart.

Swift's music is very emotional, very personal. It makes sense. Her Design Earth is in Gate 6, in a channel that is about learning through intimate sexual relationships. But where does that pressure come from, to be pouring out her emotions, trying to find clarity, through her music?

Swift has Psyche (finding perfection through love) and Cadmus (using words to create understanding) in Gate 41, meeting her North Node in Gate 30 (clarity). This brings the Root Center pressure up to her Design Earth in Gate 36 and then we have

Persephone in Gate 35. It's Persephone, the maiden kidnapped and taken to the underworld, the queen of going deep into her own underworld and mastering it, who brings all that energy to the Throat for expression.

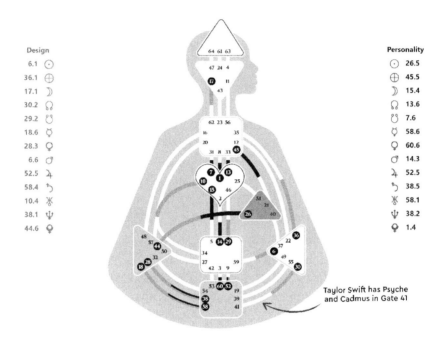

Taylor Swift has Psyche and Cadmus in Gate 41

Why did no one believe Hillary Clinton when she talked about the dangers of electing Donald Trump during the 2016 US Presidential election? Could it be because she has Kassandra (not being believed) in the same gate as her Mercury (communication)?

Why did Britney Spears struggle to get free from being controlled by her father? Could it be because she has Minerva (living by her father's rules) in a channel with her Moon (the inner child)?

Why was Freddie Mercury such an incredible singer? Could it be because he has Orpheus (music) in the same gate as his Moon (his emotions)?

Why is Serena Williams such a successful tennis player? Could it be because she has Toro (strength and perseverance) in the same gate as her Sun?

Marilyn Monroe has her Aphrodite and Lempo (the Finnish goddess of sex) in the Channel of Charisma. Her Personality Aphrodite is in the tantric Channel of Rhythm with the asteroid Photographica (being photogenic), and Mars meeting Pan (sex and fertility) in the romantic Channel of Grace. You won't see any of this in her standard HD chart! If we were looking to find Monroe's magnetic qualities, we could certainly see that she is flirting with everyone with Venus in Gate 19, but you can see how the picture unfolds with more precision once we start to look beneath the standard planets.

Prince Harry has the asteroid Lilith (women who challenge the constraints of tradition) in Gate 6 with his Personality Sun. He was never going to marry a woman who quietly fit in with the status quo! He has Heracles (also known as Hercules, the hero) in the same gate as his Mercury. In 2019 Prince Harry set up Invictus Games to help injured military recover. Heracles was considered the patron for men, particularly warriors, and used playful games as a way to recuperate.

If you feel there's more to your Human Design, you're right! And if you struggle with your strategy, or don't feel really *seen* through the Human Design lens, you might find the answer in the asteroids rather than your not-self.

———•◆•———

In this book I will be answering four questions.

1. What is an asteroid?

2. Which asteroids are most important to you?

3. How do they change the way the energy flows around your Design?

4. What new stories do they invite you to tell about your life?

It's all very well to talk about using asteroids in Human Design, but the reality is there are millions of them! There are two things people will do when they first start exploring the asteroids. They might just go into overwhelm and walk away. Or they get super excited and go on an asteroid binge but don't learn very much because they can't integrate all that information.

I want to give you a third way to approach the asteroids. I will be sharing with you the step-by-step process I have developed in more than a decade of using asteroids in my Human Design practice. When you apply this technique, you avoid the overwhelm and you can develop an embodied experience of these archetypes.

Remember, the asteroids are shards of culture held within our DNA, and they hold keys to our personal role in culture creation. At this moment in human evolution, as we are evolving and changing culture, our conscious contribution is more important than ever.

We can't do this work by learning keywords and trying to fit ourselves to them. That is the old way. We have to take ownership of the movement of these archetypes in our lives, so we can be conscious of the cultural creation we are engaging in.

———•◆•———

Let's explore what Human Design tells us about how we create culture.

The very structure of our Human Design brings four spiritual world views. The I Ching, which is the basis of the gates, comes from China. The Chakra System, which with the Kabbalah, is the basis of the Centers and Channels, comes from the Hindu tradition. The Kabbalah is a part of the Jewish mystic tradition. Then there are the planets, which create the definition in your Design. They come out of the tradition of western astrology.

It is only in the astrology that we find anything of our western spiritual tradition.

Carl Jung made a fascinating point in his book *The Archetypes and Collective Unconscious*. He argued that those of us in the western world became cut off from our heritage when it was overtaken by Christianity. Other than some deep-rooted pagan practices and beliefs that refused to die, we lost access to the deeper creative and spiritual life of the indigenous cultures of Old Europe. Now, in modern times, once we reject established Christian religion, we have little option but to flock to eastern religions to find reflections of our spiritual lives. Jung believed this has led to an impoverishment of our own spiritual heritage.

We can replenish our spiritual roots if we dig deeper into the original versions of many of these archetypes from Ancient Greece and Rome.

I learned this from Pandora.

Most of us know that Pandora had that famous jar. She was too curious, opened the lid, and oops, out came pestilence and some other nasties. Luckily the silly girl managed to get the lid back on before hope escaped!

This version of Pandora was the creation of the ancient Greek poet Hesiod in *Theogony*. Hesiod described how Zeus was angry with Prometheus for stealing fire from the gods and gifting it to humanity. Zeus hatched a plan to give humans some of his own gifts. He commanded Hephaestus to create the first woman, whose descendants would torment the human race.

This is Hesiod's Theogony[2]

> *For from her is the race of women and female kind: of her is the deadly race and tribe of women who live amongst mortal men to their great trouble …*

But there's more. In a later poem "Works and Days"[3], we discover that

> *…all who dwelt on Olympus gave each a gift, a plague to men who eat bread.*

The gods of Olympus used Pandora as a kind of vessel to load "gifts" that were actually curses onto humanity. Hesiod closes with this admonishment[4] "there is no way to escape the will of Zeus" In other words: Humans, you stole my fire, and you got what was coming to you.

---

[2] http://www.perseus.tufts.edu/hopper/text?doc=Perseus%3Atext%3A1999.01.0130%3Acard%3D585

[3] http://www.perseus.tufts.edu/hopper/text?doc=Perseus%3Atext%3A1999.01.0132%3Acard%3D59

[4] https://www.perseus.tufts.edu/hopper/text?doc=Perseus%3Atext%3A1999.01.0132%3Acard%3D83

By examining the story of Pandora, and her representations on pottery and elsewhere, archaeologist Jane Ellen Harrison observes how a much earlier version of the goddess, known as all-gifted (pan-dora) devolved over time into a mortal woman foolishly propagating curse. The life-giving goddess is diminished to become a death-giving human woman.

Not only did Pandora get me thinking about the mythologies surrounding the archetypal figures in Ancient Greece and Rome, but she also taught me to question how astrologers were interpreting them.

The most common interpretation for the asteroid Pandora is that she shows a place in your life where you are too curious, leading to unexpected disasters. This seems to me to be a very shallow take. It was Pandora who encouraged me to do my own discovery charts and research before I started working with an asteroid. Pandora, like all the asteroids I have researched, has a very synchronous discovery position, Gate 21, which is about biting through the surface to get to the deeper truth. In my work with Pandora, I consider her an indicator of where we have learned to believe that our true gifts are actually curses that we must keep contained. Hold tight to that lid! The gate in which we find Pandora is one in which we are gifted and giving, but where we have been led to believe that sharing our gifts will harm people.

By getting stuck in the superficial narrative about these archetypes we lose touch with the depth and richness of our own lives, seeing our own experiences only through the lens of a cultural movement intent on drawing power from the collective and consolidating it in their own hands. We have lost touch with the very means to the stories that will support our evolution.

Using Asteroids in Human Design

What are Asteroids?

Asteroids are bits of rock floating in space, orbiting around our Sun. They don't qualify as planets because they are too small and don't have their own gravitational field.

Like planets, asteroids are interesting to both astronomers and astrologers. Astronomers love asteroids because they give clues to the formation of the universe. Astrologers use them because they give clues to our psyche.

There are millions of asteroids in our solar system, but only a few thousand have any significance in Human Design. And only a handful of those will be important in your life.

Where Do We Find Asteroids?

The majority of asteroids orbit in what's known as The Main Asteroid Belt, in the area of our solar system between Mars and Jupiter.

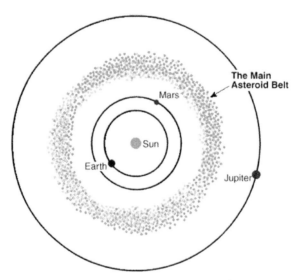

Orbits drawn approximately to scale[5]

From an astrological point of view, their position in the solar system is significant. Astrologer Demetra George describes them as a psychological bridge between the planets representing your inner nature—Mercury, Venus and Mars, and the realm of culture and society—Jupiter and Saturn. This holds true as much in Human Design as it does in astrology.

———•◆•———

You've probably heard the names of some of the most important asteroids, like Ceres and Psyche. The name of an asteroid is important. Just like we know the influence of the planet Venus is different from Mars because of the name, we can tell something about the influence of an asteroid in the same way.

---

[5] http://science.nasa.gov/science-news/science-at-nasa/2000/ast17apr_3/

The names of asteroids matter to us. Zeus will always have a much deeper resonance for us than, say, Gaskin. Zeus is part of our cultural substructure, our collective consciousness, heroic, invincible, irresistible. The asteroid Gaskin was named after Regina Gaskin, a mentor in the 2003 Intel Science Talent Search. Which is lovely, but not so powerful as a collective archetype.

Astronomers use naming conventions created by the International Astronomical Union. Originally the planets were given the names of male gods from Greek and Roman mythology. Main belt asteroids, which have regular orbits, were named for the Greek and Roman goddesses. Asteroids with unusual orbits, like the Mars-crossing asteroid Eros, were given male names. Yes, that's right. The well-behaved asteroids were named after the goddesses and the adventurous asteroids after the gods!

These days, because of the sheer number of asteroids, names are no longer limited to mythology. Historical characters like the economist Adam Smith, eminent scientists like Einstein, and even pets, are found in the list of named asteroids.

These names are only considered standard within the western scientific tradition. Most indigenous cultures had their own names (and mythologies) for planets, stars, and visible asteroids, like Ceres and Vesta.

The focus on names from Greek and Roman mythology mirrors their importance as building blocks for western culture. This is why these astrological bodies are so potent for us, as fragments of culture we hold within our individual psyches.

There is often both a Greek and a Roman version of the planets and asteroids, Aphrodite and Venus, Poseidon and Neptune, Artemis and Diana, Proserpina and Persephone for example. They have different meanings, and greater or lesser significance, often because of their size. A planet is more important than an asteroid. Pallas is more important than her Greek counterpart Athene because she is a much larger and more prominent asteroid.

I'm disappointed there are almost no asteroids named for Celtic, Welsh, and Gaelic cultures. Even major figures such as Brigid and Boudicca are not found in the named asteroid list. Access to a personal relationship with these Celtic mythologies may have given us a more robust base from which to recover our feminine DNA.

As the Roman orator Marcus Borealis said:

> *The women of the Celtic tribes are bigger and stronger than our Roman women. This is more likely due to their natures as well as their peculiar fondness for all things martial and robust. The flaxen haired maidens of the north are trained in sports and war while our gentle ladies are content to do their womanly duties and thus are less powerful than most young girls from Gaul and the hinterlands.* [6]

The most powerful work we can do with these Greek and Roman gentle ladies content with their feminine duties is to uncover their original nature, back before they were domesticated.

Some astrologers make a clear distinction between the Greek and Roman archetypes. There are some differences in how we interpret two asteroids like the Greek Hera and the Roman Juno. They have different discovery charts and the asteroids themselves vary. For example, Juno is the fourth named asteroid and significantly larger than Hera. Although Hera is the more dominant mythologically and can add to our understanding of the meaning of Juno, the Roman asteroid is more important in your Design.

I don't always distinguish between the Greek and Roman when I'm telling the stories of the myths. This is because I give first importance to the planet and asteroid. For example, I may talk about Ceres and Persephone, even though Ceres is Roman and Persephone is Greek. I do this because these are the two most significant representative archetypes. There are asteroids called Demeter and Proserpina, but they carry less significance. It's not to say you can't go and explore their position in your Design though!

———•◆•———

It's not just the name, but also the timing of the discovery of an asteroid that makes a difference in its meaning. When I'm

---

[6] https://link.medium.com/ni17vFYmEfb

researching an asteroid, I start by creating a chart for the moment it was discovered—its own personal Human Design!

You will notice as you go through the section where I talk about each individual asteroid that a lot of them were discovered in Gate 3, Difficulty at the Beginning. This suggests to me that the asteroids are here to help us navigate difficult times at the beginning of a new phase of human consciousness.

When an asteroid or planet is discovered, it seems to have an effect on us, as if its energy suddenly bursts forth into our lives. Uranus, the planet of revolution, was discovered in 1781 amidst the revolutionary beginnings of the French and American republics. Neptune was discovered in 1846, just as the nature of spirituality was changing, with theosophy blending the religious beliefs of east and west.

The early 1800s, when so many asteroids were being discovered and named for versions of the goddess, came to be known as the first wave of feminism. Women's role in society was changing, particularly with the focus on the suffragettes and women's right to vote.

In the 1970s, during the second wave of feminism, which was focused on sexual freedom and reproductive rights, astrologer Eleanor Bach published the first asteroid ephemeris. A few astrologers like Martha Lang-Wescott began using asteroids at this time, but most thought they weren't worth all the extra effort.

This changed in 2005 when astronomers discovered some new planets way out at the edge of our solar system beyond Pluto. Suddenly it was impossible to justify limiting ourselves to the standard pantheon of planets. For me at the time, shifting my view of Human Design to include both asteroids and the new outer planets posed a question—is this still Human Design?

——•♦•——

I need help with how to frame this information.

When I started exploring the asteroids, I used an online astrological ephemeris. I plugged in the person's birth details and got a list of asteroids with their astrological position. I then, painstakingly, went through that list figuring out what gate each asteroid was in, using the Human Design Index.

In 2015, we created Taraka, the online Human Design software that does all that hard work of calculating for you.

The Meaning of the Asteroids

When I prepare for a session I have a process I follow and in this chapter I'm going to share that with you. But first, a bit of wise advice. As we know, there are hundreds of thousands of asteroids. I suggest you follow the process in this book as a way to map things out and get an overview of your most significant asteroids. Even then, you will have maybe as many as twenty significant asteroids, and that is a LOT of information. Don't try to understand it all at once.

You will find over time that your interest in particular asteroids will come and go. That is because at any given time, some asteroids are more active in your life because of transits and the movements of your Holographic Human Design chart layer.

You may want to start by exploring, let's say, Artemis even though she is on your Mercury and not your Sun. If you look at both the transits and your Holographic layer you may find that gate is really active, which means Artemis is alive for you at that particular time. Take advantage of those "alive" moments when the archetypal energy is calling you to engage!

You should also pay attention to any asteroids that repel you. They are likely to have a message that you are resisting or want to draw attention to aspects of your life you would rather avoid.

I often talk about an asteroid being prominent or significant in a Design chart. What does this mean? As you read through this chapter, you'll learn about what makes an asteroid important to you. I have Apollo in the same gate and line as my Personality Sun. That is very significant! An asteroid in the same gate as my Personality Jupiter, for example, will play out somewhat, but. It's definitely not prominent in my life.

Personality and Design Layers

When I first started using asteroids, I focused on the Personality layer in the Human Design chart. I figured the Design layer was operating under conscious awareness and we probably wouldn't notice the influence of the archetypal energy there. It turns out that's absolutely *not* the case! And it might even be that the asteroids in the Design layer have a more powerful influence in our lives.

When people are struggling, the issue will almost always surface in the Design layer asteroids. I think this is because these asteroids operate outside of our conscious awareness, but their effects are very real in our lives. We have problems but we can't quite grasp consciously what they are or how to deal with them. This is cultural conditioning; we are playing our approved cultural and family role and find it difficult to see beyond that to find a new emergent solution. Our more conscious expression of the archetype can be found in the Personality layer.

For thirteen years Britney Spears struggled under the control of her father who set up a conservatorship, a form of legal and financial guardianship, after she had some very public mental health crises.

Spears has Minerva in the Design layer of Gate 19 with her Personality Moon. Gate 19 is about the material and emotional support we get from our family. The Personality Moon shows what we need to feel secure, in this case she would struggle with fears of not being supported. Because this is a Personality activation, it's likely Spears knows this is an ongoing fear in her life.

The asteroid Minerva, however, sits in the Design layer, which means she is not consciously aware of it. Minerva can indicate places in our lives where we follow our father's rules, either through unconscious family belief patterning or, as in this case, cultural or legal requirements. Even though it's in the Design layer, Minerva has obviously played a huge part in Britney's life in the last decade. Working with that energy more consciously will bring her a different kind of security than she had previously imagined. Minerva is a Greek version of the Roman Pallas, and in her healthy expression she is a gifted strategist and artist.

Why did it take so long for Britney Spears to figure this out? It's all in the timing of the transits and the Holographic Human Design layer.

In November 2021, when Spears was seeking to end the conservatorship, Saturn (self-responsibility) and Jupiter (new opportunities) were transiting in this same channel with her Moon and Minerva. At the same time, her Holographic Mercury (what needs her conscious attention) went into Gate 19. It was through the transits and shifting Holographic layer that issues sitting under her conscious awareness came to the surface to be dealt with.

It's important to understand the beauty of timing in this process. If she had tried to resolve the issue before this, it's likely she wouldn't have had the awareness to get the best outcome. This is one of the underlying processes supporting the idea of waiting in Human Design—the perfect time will always come. It's also why I suggest when studying your asteroids, that you go with what is drawing your attention. There is an underlying universal harmony that is more powerful than your idea about what you should be learning next.

———•◆•———

It's time to create your map that will give you an overview of your most important asteroids.

Defined Gates

We start by looking at your defined gates, and that means also looking at the planets. Some of you won't be accustomed to looking at which planet is in a gate, but when we are exploring asteroids, it matters.

There are four layers of planets.

Firstly, your Incarnation Cross, which is made of your Sun and Earth activations. I include the Moon in this first step. When I add the Moon to the Sun and Earth, I call it the Incarnation Matrix. Then we have the inner planets—Mercury, Venus, and Mars—which make up aspects of our personality. Next, we have Jupiter and Saturn, the planets that rule over our relationship with culture and society. Finally, we have the outer or transpersonal planets, Uranus, Neptune, and Pluto. We also need to include the Nodes, which are not planets at all, but show our life geometry, where we've come from and where we are going in our lives.

White Gates

After exploring the asteroids in our defined gates it's time to move on to the white gates.

Just like the planets, not all white gates are the same. The most important white gates are your bridge gates, creating a bridge between areas of definition. Any asteroids in these bridge gates will influence different areas of your design, let's say emotions and mind, relate to each other. If you have a single definition, you don't have any bridge gates.

The next most significant white gates are called hanging gates, they don't create a bridge between areas of definition, but they still connect with a planet. Britney Spears has a single definition, so she doesn't have any bridge gates, but she has a lot of hanging gates. Notice that she has all three Head Center gates. The three Ajna

Gates—47, 24, and—are all hanging gates and give her inspired mind access to the Ajna Center. She has the asteroid Ceres in Gate 47 making a channel with her Design Sun in Gate 64, which is very significant.

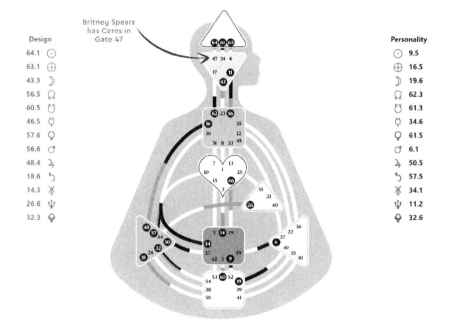

Spears has her Personality Sun in Gate 9 and Gate 52 is a hanging gate, helping her ground her personality through the Root Center. She has Circe (transformation) in Gate 52.

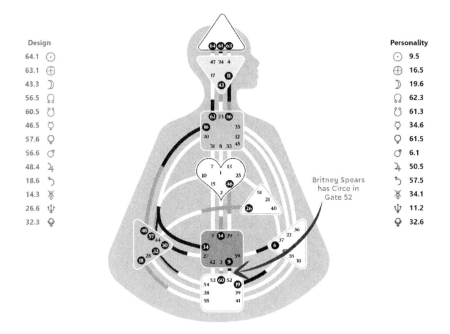

White Channels

The last area we want to explore in our process is the white channels, which most people ignore in their Design. Looking again at Britney Spears's Design, she has Aphrodite in Gate 38 (Opposition) and Juno (equality in relationships) in Gate 28 (Great Exceeding). These are both powerful archetypal energies in a channel that is seeking to find ways to connect with others without losing our individuality, and that needs life to be a great adventure rather than a struggle. So don't ignore the white channels!

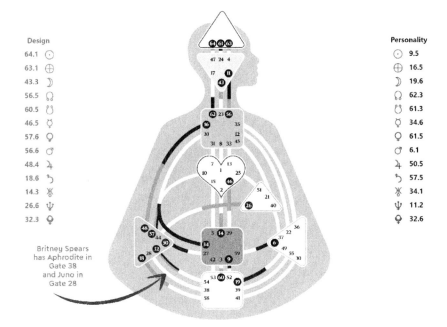

Design

64.1 ☉
63.1 ⊕
43.3 ☽
56.5 ☊
60.5 ☋
46.5 ☿
57.6 ♀
56.6 ♂
48.4 ♃
18.6 ♄
14.3 ♅
26.6 ♆
32.3 ♀

Britney Spears
has Aphrodite in
Gate 38
and Juno in
Gate 28

Personality

☉ 9.5
⊕ 16.5
☽ 19.6
☊ 62.3
☋ 61.3
☿ 34.6
♀ 61.5
♂ 6.1
♃ 50.5
♄ 57.5
♅ 34.1
♆ 11.2
♀ 32.6

Incarnation Cross

I always start my examination of the asteroid layer by looking at the Incarnation Cross.

The Incarnation Cross is made up of four activations in your Design—the Personality Sun and Earth, and the Design Sun and Earth. It's called a cross because those four points do actually form a cross on your Human Design mandala. It looks like this:

It takes a lifetime to bring each arm of that cross into relationship with the other three. They don't naturally flow together as a team. I describe it as trying to mix oil with water and it takes a lifetime to learn how to shift their relationship from opposites to a complementary partnership. Asteroids in those four Incarnation Cross gates are going to form a significant part of our personality. We may be consciously aware of their influence, but if they are in the Design layer we may not.

Britney Spears has her Personality Sun in Gate 9 Line 5. She does best when she takes the time to cultivate her life by focusing on small details. She also has the asteroid Metis (wisdom) in Gate 9 Line 5, but it's in the Design layer. This makes sense if we think about how her Minerva (strategy, artistry, pleasing the father) was also in the Design layer with her Personality Moon. There's a theme here of not realising her own capacity to be wise and create successful strategies for her own life.

Prince Harry has Atropos, one of the three Fates, with his Sun in Gate 6. This is fascinating because his mother Princess Diana also had Atropos in the same gate as her Sun. Atropos was one of the three Fates, the one who brings things to its inevitable conclusion. It makes me think about how much his mother's death has influenced Harry, and perhaps they are working across dimensions to bring the Royal Family to an end.

Just like all the planets, the quality of the Sun and Earth will affect the role of the asteroid in your life. The Sun will bind the archetypal energy of the asteroid into your personality, sometimes making it difficult to even discern as a separate energy. The Earth will make it hard to reach, as if you have to find a way to allow it, to embody the archetypal energy.

Personality Sun: Diversity, maintaining your own individuality while accepting the difference of others.

Personality Earth: Connectedness, balancing the individualism of the Sun by connecting with others who support you to express yourself in healthy and creative ways

Design Sun: Interaction, the energy and spirit that powers your individuality, facilitating the flow of living universal knowledge

Design Earth: Adaption, able to be transformed by interaction without losing your own anchored embodiment.

I look at the Moon at the same time as the Incarnation Cross, in a format that I call the Incarnation Matrix. Ra Uru Hu says the Sun carries 70 percent of the neutrino stream and is the most important activation in your Design, however much of astrology was based more on the importance of the Moon over the Sun. The Moon can represent the very young child within us, and what it is we need to feel happy, safe, and secure.

The Sun, Earth, and Moon are constantly interacting with each other, this is how the New and Full Moons each month come about. When I'm looking at the significance of the asteroids, I believe the Moon has as much significance as the Sun and Earth.

We've already looked at Britney Spears Personality Moon activation, and the asteroid Minerva. If we look at her Design Moon in Gate 43, we find Persephone. Her Design Moon is about taking a stand to be her authentic self. If we bring Persephone into the mix, we can see the potential for her to be 'kidnapped' and suddenly find herself in places where she feels out of her depth. Persephone tells us that Spears gets a sense of security from being constantly on the alert for the unexpected.

Dolly Parton is known for keeping her marriage out of the spotlight, but she and Carl Dean have been married since 1966. Parton has her Moon in Gate 59, which tells us, amongst other things, that she gets her emotional security from intimate relationships. She has the asteroid Hera in the same gate, the asteroid most associated with being a wife. Hera was much more interested in her husband than her children, and Parton and Dean don't have any children.

The quality of the Moon in your Design shows how we find emotional stability and sustenance through cycles of experience. Asteroids in the same gates as your Moon activations will give you a deeper understanding of your emotional needs.

Personal Planets

After the Incarnation Cross, we can move on to explore the personal planets Mercury, Venus, and Mars.

These planets are archetypal aspects of our personality. They form the story of how we have developed. Mercury represents the curious young child and what she loves to explore, think, and communicate. Venus is our budding young need to be valued in our relationships, to be loved, to be included. In developmental terms, Mars kicks in during the terrible twos, when we want to express our own will, act on our desires, and find satisfying ways to channel our creative energy.

I always take into account these development stages when exploring the asteroids. It's likely that Britney Spears developed the strategy of pleasing her father at a very young age, the Moon tells us that. We can also see that as she got to around age two and was seeking more independence through expressing her Mars energy, Gate 6 (Conflict), where she is trying to argue or perhaps avoiding arguments within the household. She has Hebe in that gate, the handmaiden and helper who sees how she could help others, if only they would stop arguing with her!

Exploring the Design layer asteroids can give us important insights into how we respond to situations in our lives. This is a powerful way to use the asteroids to help us see unconscious aspects of our personality and situations we may have encountered when we were too young to remember their influence.

Spears has Terpsichore (the muse of dancing) in the Personality layer of Gate 6 with her Mars. She would be conscious of her love of dance. In the Design layer she has Klotho (knowing when enough is enough) in a line that suggests she stop arguing and walk away. Klotho could also suggest a fear of walking away from the argument (she might die, Klotho is one of the Fates who give us our life allotment).

While these personal planets are the building blocks of our personality, they also represent aspects of our adult self. The asteroids we find in the gates with these planets tell us a great deal about our fundamental ways of being in the world.

We can see for instance, how Hebe and Harmonia continue to express themselves through Parton's quiet and unassuming charity work.

Dolly Parton has spoken often about her difficult childhood. She experienced extreme poverty, and her family often went without food. She has the Gate of Limitation (60), activated by her Sun (identity) and Venus (the finer things in life). She also has asteroid Hebe in that gate. Hebe's job on Mount Olympus was to serve the gods food and drink, they were never to go without! I can imagine young Dolly, with that Hebe forming part of her personality, feeling a personal sense of responsibility for not having food on the table. We can see how much this experience formed part of her sense of self (Sun).

Freddie Mercury would have learned early to use his capacity to communicate in powerful ways. I think of him working those massive stadium crowds in songs like "We Will Rock You," breaking through the personal barriers of each individual and getting them to be a part of something bigger.

He has his planet Mercury in Gate 59, which has a strong connection with sex, and it is joined by Sappho (the poet). His stage presence communicated a mix of poetry and sex, and his communication could feel intrusive to others. With Alethia (truth) in the adjoining Gate 6, he would have been a difficult child, making others uncomfortable by penetrating their protective psychological barriers. He could have become quite silent to protect himself from the way this tendency impacted others, but instead, he was gifted with courage (Mars), self-belief (Jupiter), and instinct for self-preservation (in Gate 57) to know when it was safe for him to be outrageous.

The planet Mercury represents our intelligence, communication, what we love to talk about, and the process by which we bring ideas to consciousness.

Venus represents our desire for love, beauty, and relationship. She delights in the sensuality of existence and the pleasure of beautifying it.

Mars represents how we initiate the raw life energy we use to chase our desires and passions, to fulfill our needs, and how we use assertion or aggression to protect ourselves.

<hr>

When we look at Jupiter and Saturn, we see the building blocks of culture and society. At the simplest level, Jupiter is opportunity and Saturn is how we work to take advantage of that opportunity. Saturn is how we organise and govern ourselves and Jupiter rules over law and religion. On a cultural level, our Saturn shows our relationship with authority and Jupiter our relationship to law. On a personal level, Jupiter is expansion and Saturn can operate as a bridge between our soul purpose and the creation of an individual authentic self, or a wall we build out of fear of not being enough. Saturn is what we choose to take responsibility for, and we can get trapped in taking responsibility that leads us into dead ends, surrounded by rigid walls of old responsibility that we forgot to take down. Saturn is also our relationship to authority and self-authority.

Asteroids in the same gate as these two planets can tell us more about how we contribute to society.

Princess Diana has Narcissus on both her Saturn and her Jupiter. Her Personality Narcissus is in Gate 60 with her Saturn, and her Design Narcissus is in Gate 41 with her Jupiter. The asteroid Narcissus shows where we have healthy ego needs that don't get met (we all need a healthy amount of narcissism to develop self-love). Narcissus can also show where in our design we can draw lessons on how to love our true self and not just the reflection we receive from others of who we are.

We can see how Diana was limited in her ability to achieve authority in her own life when she became a part of the Royal Family. Jupiter gives a clue about how she can expand beyond Saturn's rigidity, and what she truly needs to take responsibility for. Having her royal title removed was her opportunity. She took control of her own image and became a reflection of self-authority for others and had a profound effect on society.

When I'm working with Narcissus in a chart, I always look at the asteroid Echo. Echo shows where we keep trying to break free of that projected reflection of who we are. Diana's Personality Echo is in Gate 15, on her Personality Mercury! When she began to find her voice (Mercury) she learned to work with her own extremes and eventually find some integrity. Her Design Echo is in Gate 27 (nourishment) which explains why she used bulimia to try to draw attention to her needs.

John Lennon has his Design Jupiter and both his Saturn activations in Gate 24. He also has the asteroid Astraea there. Astraea is the Star Goddess who, according to myth, was the last immortal to abandon humanity during the last Iron Age when life became too hard for her sensitive soul. As she left, Astraea assured humanity that she would be the first to return at the dawn of the next Golden Age. I find it interesting that Lennon has Pandora in Gate 61, an asteroid that can indicate hope when all is lost. This is a channel of imagination, and Lennon helped us hold onto hope that a better world is possible.

Jupiter is where you are naturally expansive, gregarious, generous, and seeking to join in the fun of life. It's where we can find our faith and confidence in our life's path.

Saturn represents the principles of limitations and boundaries, it can be a bridge or a rigid wall of fear of being not enough, a bridge between our soul and creation of our individual authentic self, development of wisdom over time, what we choose to take responsibility for and our relationship to authority and self-authority.

■•◆•■

Uranus, Neptune, and Pluto are known as the transpersonal planets. They take years to move through a gate, so you will share that particular activation with many others born around the same years as you. The outer planets often feel like energy that comes from outside us, and it can take a lifetime to realise how we express it, and to integrate these powerful forces into our personality. The asteroids we find in the same gates as the outer planets are not going to play out as personally in our lives as they do with the faster moving planets.

If we have a personal planet in the same gate as a transpersonal planet, we have a personal filter through which that outer planetary energy will express. It's the same with the asteroids, which show ways for us to integrate the larger more impersonal forces of Uranus, Neptune, and Pluto.

Oprah Winfrey has Uranus in Gate 53 and her Child asteroid is also there. In this gate you say yes to something new and then you have to grow into it. With Uranus there, Oprah's childhood would have brought unexpected events. In this gate she had to find her unique (Uranus) way of responding.

Beyoncé has the asteroid Echo in Gate 14 with her Uranus. This is about knowing how to apply her skills and talents in the right direction. Her Uranus activations in Lines 3 and 4 are about having visionary qualities, being patient, and taking time to grow into her gifts. In Line 6, Echo shows her personal experience of that energy, telling people over and over again that, despite her freakish talent (Uranus in the individual circuitry) she knows she will be famous, that everything is unfolding just as it should.

Actress Kate Winslet has Magdalena in the same gate as her Uranus. In Gate 50, this is about making the world a better and more caring place by being a unique individual who expresses the transpersonal energy of Uranus through a personal experience of the divine feminine. One of the aspects of Uranus is androgyny, and Winslet has her Uranus in a channel with her Mars (the masculine). The combination of the two is apparent in many of Winslet's roles where she holds a deeply feminine energy regardless of how tough and uncompromising her character.

Lead singer of INXS, Michael Hutchence, is a great example of the adulation Neptunian personalities experience. He has Design

Neptune in Gate 28 with his Design Sun, Design Mars, and Personality Moon. People with a strong Neptune influence in their Design can be very sensitive and feel easily overwhelmed. He also has Aphrodite in that same gate, meaning relationships were his primary way of staying grounded. Neptune doesn't find life easy; it can feel harsh and people with a strong Neptune will seek escape through addictions like relationships or drugs. In the documentary *Mystify*, people who knew Michael say repeatedly that his life was all about experiencing pleasure (Aphrodite). He found it very difficult to be alone, needing to have his Aphrodite by his side. But also, like Aphrodite, he didn't allow the rules to quash his desire for relationship, beauty, and play as the prime creative forces of life. In a sense, that Aphrodite is what made life real for him.

John Lennon was another Neptunian character, with Neptune and Design Moon in Gate 6, meeting Venus in Gate 59. He had Urania in Gate 6, giving him a genius at symbolic language where a picture paints a thousand words as in his many lyrics (i.e., merry-go-round, nowhere man/land). This was likely a way for him to manage all that conflict and friction (Gate 6) when all his planets there are so sensitive to anything that's not harmonious, finding that one perfect phrase that brings the elements into harmony.

Queen Elizabeth has her Pluto (power) in Gate 39 (Limping). She has no asteroids in that gate indicating personal conflicts between her sense of personal power and external power, but we do see in her multidimensional chart layers her struggle with having to put her own life aside to take on the role of monarch. Prince Philip, however, has the asteroid Karma in the same gate as his Pluto, showing a conflict with having to own his karma and find his peace in playing second fiddle to his powerful wife.

———— • ◆ • ————

Unlike the rest of the planets in your Human Design chart, the Nodes don't actually exist. They are what is called in astrology a *calculated point,* and the point they are calculated from is your Moon. (Every planet has Nodes by the way!)

Along with the Incarnation Cross, the Nodes make up your life geometry. They represent your movement from your past (South Node) to your future (North Node). We tend to feel comfortable in our South Node gate and uncomfortable in our North Node gate. In a sense, everything else in your Design is there to support your North Node activations.

So what can the asteroids tell us?

I was curious to see what Beyoncé's Design might reveal, because she was famous from such a young age, and her first band was called Destiny's Child! It's as if her North Node immediately took hold of her! She has a minor planet called Orcus in the same gate and line—Gate 56 Line 5—showing a great life mission. But she also has asteroids there that tell us something about her life geometry. Arachne represents a very high level of skill and learning to apply that skill without being too proud. Beyoncé's South Node in Gate 60 Line 5, Limitations, suggests she is comfortable breaking out of situations that are too small for her. Medea is there too, with a natural gift for finding solutions to tricky problems.

The South Node is an energy we have to resist falling back on. Medea gives us a clue about that. If you have limitations, and a gift for working through them, it's possible you move too fast to connect with the people who recognise your gift (North Node), or you may come across as proud and pushy (Arachne).

<hr>

So far, we've talked about asteroids in already activated gates and how they add a flavour to the quality of the gate and planet. But what about all the white gates in your Design? This is where things get really interesting.

You might not have paid much attention to those white gates. Some will have no influence on your life at all, others will contain some of your most important experiences.

The most important are the bridge gates—the gates which bridge any separate areas of definition you have. Bridge gates connect aspects of yourself that don't usually have access to each other. If you have a split definition, you might have your emotional and mental areas separated or your splenic instinct may be connected to your Root Center, but your mind is connected to your Throat, which means you are not always mentally aware or able to express your instincts. The gates that make these connections are some of the most important aspects of your Design.

You can see how Queen Elizabeth has two separate areas of definition—mental (Head and Ajna) and emotional (Solar Plexus and Root). There is one particular gate that bridges the split—her Gate 35.

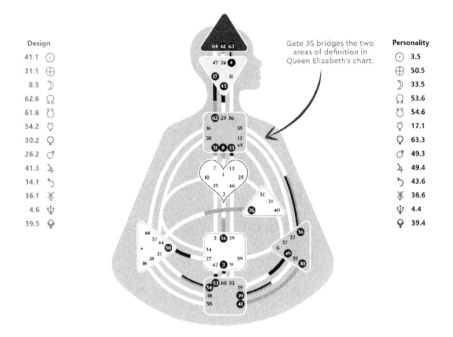

Gate 35 bridges the two areas of definition in Queen Elizabeth's chart.

The next most important white gates are the so-called hanging gates. These are the white gates adjoining an activated gate to make up a full channel. The most important hanging gates will connect an awareness or motor center to the two manifesting centers, the throat and root.

You can see in Dolly Parton's Design that Gate 16 is white, but any asteroids in that gate would influence the flow of instinct from Gate 48 to her Throat Center for action and expression. Parton has no significant asteroids in Gate 16. I find that when this happens, we truly do experience the need for others to bridge that energy for us. Not only is there no definition in that gate, but there is also no subtle minor planet there either. We are very open to the influence of others in these gates.

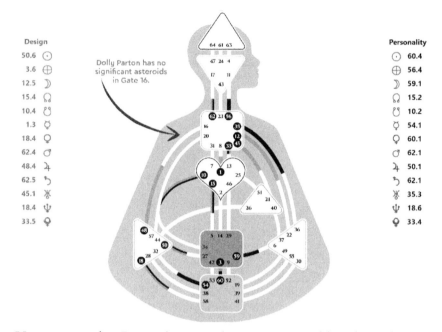

Design

50.6 ☉
3.6 ⊕
12.5 ☽
15.4 ☊
10.4 ☋
1.3 ☿
18.4 ♀
62.4 ♂
48.4 ♃
62.5 ♄
45.1 ♅
18.4 ♆
33.5 ♇

Dolly Parton has no significant asteroids in Gate 16.

Personality

☉ 60.4
⊕ 56.4
☽ 59.1
☊ 15.2
☋ 10.2
☿ 54.1
♀ 60.1
♂ 62.1
♃ 50.1
♄ 62.1
♅ 35.3
♆ 18.6
♇ 33.4

You can see that Parton has two throat gates reaching down for her Solar Plexus Center and connecting those would give Parton emotional access to the Throat. She has Aphrodite in Gate 22, for

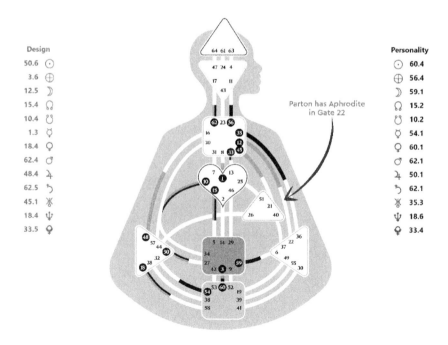

Design

50.6 ☉
3.6 ⊕
12.5 ☽
15.4 ☊
10.4 ☋
1.3 ☿
18.4 ♀
62.4 ♂
48.4 ♃
62.5 ♄
45.1 ♅
18.4 ♆
33.5 ♇

Parton has Aphrodite in Gate 22

Personality

☉ 60.4
⊕ 56.4
☽ 59.1
☊ 15.2
☋ 10.2
☿ 54.1
♀ 60.1
♂ 62.1
♃ 50.1
♄ 62.1
♅ 35.3
♆ 18.6
♇ 33.4

example, looking to get to the Throat via that Design Moon in Gate 12.

Lastly, there are combinations of gates that bring some aspect of your Design into a connection with other parts of you.

I often work with people who have important personal planets in an undefined Head Center with no way to connect or express. Physicist Werner Heisenberg won the Nobel Prize for the creation of quantum mechanics. He was also a fan of Fritjof Capra's exploration of the links between modern physics and eastern mysticism.

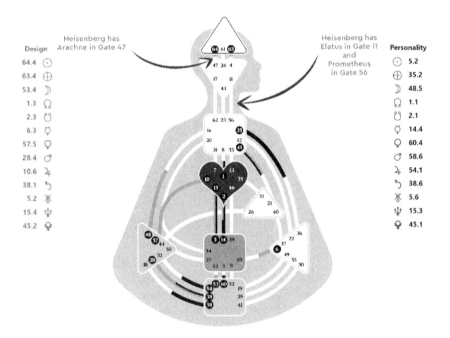

His Design Sun and Earth are in the Head Center in Gates 64 and 63. These are gates of mental innovation, bringing through new information to support change. It can be helpful to know what asteroids are in the large gap between those two gates and the Throat Center. Heisenberg has Arachne (brilliance, making new connections) in Gate 47, Elatus (brilliant communication) in Gate 11 and Prometheus (new consciousness tools) in Gate 56. Although Heisenberg would still need people around him to support the flow of inspiration from his Head Center, we can see the subtle pathways of his mind by exploring the asteroids in those white gates.

White Channels

Even white channels can be significant. I have nothing activated in the emotional manifesting channel made up of Gates 36 and 35, but I do have Pallas in 36 and Prometheus in 35. My Pallas (creative brilliance) is constantly working towards finding the wounded light in humanity (Prometheus) in order to release humanity from the clutches of the gods. If you follow my blog, you'll know that's a strong theme. Yet, if I was only looking at my standard Human Design chart that wouldn't be obvious at all.

Most people would describe Sir David Attenborough as charming. It's a quality I associate with the Channel of Grace, Gates 22 and 12. But that channel is white in Attenborough's Design. Unless we look at the asteroids layer, where we find he has Cadmus (an ability to explain things and educate) in Gate 22 and Arachne (high level of skill) in Gate 12. This is a channel that influences others through the tone of voice and can attract a great deal of attention.

Asteroids in white channels do not give definition, but they do bring their influence to the flow of energy through that channel, and a greater understanding of the subtle influences guiding our lives.

——•◆•——

Now that we know where to look for asteroids in your Design, it's time to explore what they mean. You might like to have a place you can go to get a simple explanation for each asteroid, but it doesn't work like that, for a number of reasons.

Firstly, people will always put their own interpretation on an archetypal energy. If we're looking at Psyche, for example, you might be most influenced by the part of the myth where her father gave her up for marriage with what everyone thought was a terrible monster (but later turned out to be the Eros, the God of Love). Or maybe you resonate more with her challenging relationship with mother-in-law Aphrodite. Or perhaps you are inspired by her willingness to go through so many difficult trials for the sake of true love.

Looking at the mythology is an important component in understanding the meaning of an asteroid. But there's another lens through which we can explore. We can also learn something about the meaning of an asteroid by creating a Human Design event chart for the moment it was discovered. It is, in a way, the asteroid's personal Human Design chart.

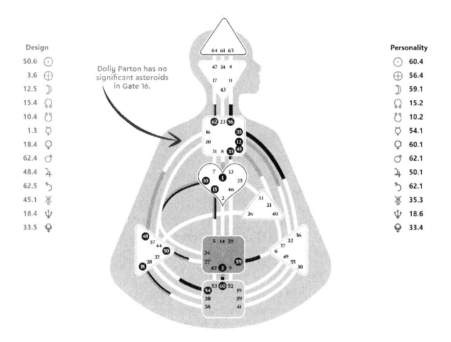

If we look at the discovery chart for Psyche, for example, we can see that Psyche is in Gate 29 Line 3 which describes a difficult situation and suggests you wait to get the full measure of it before you act. In *The Gene Keys[7]*, it is described as leaping into the void and devotion.

I don't just look at the position of the asteroid but also at the other planets. In this case, we find Atlantis (a feeling of doom) also in

___
[7] ISBN 978-1-780285429
Richard Rudd, *The Gene Keys: Embracing Your Higher Purpose.*

Gate 29 and Arnica (a treatment for shock) in Gate 46. The entire channel is about committing to something that may be difficult but is absolutely necessary for your success. Saying yes to love, not being too quick to judge the other person, feeling overwhelmed by warnings of doom which make us cautious, and then Arnica comes to the rescue in the gate of actively seeking the warmth and light, creating a sensitivity and fear that has us isolating ourselves from others.

Looking at the Sun and Earth we see the message that we should hold true to our own convictions even in difficult times. The Moon in the Gate of Revolution suggests having confidence in your own intuition and eventually you will bring others to your side.

This informs how we work with the asteroid Psyche in our Design. She represents a transformation, a revolution, brought about by our ability to hold true to love. If we are not overwhelmed by dire warnings (think of Psyche's sisters whispering to her that Eros must be a monster!) and proceed cautiously in line with our own intuition, we will overcome our own fears and succeed. The gate in which we find Psyche in our Design helps us understand the gifts we bring to that process.

Princess Diana's Psyche in Gate 62 suggests she needed to take great care when committing to love. Each step tested the relationship, as she gathered facts to support her intuition. Was this the type of love she needed? She had Ceres, Aphrodite, and Child in Gate 17. This tells us how important romantic love was to her, Ceres is seeking nourishment, Aphrodite beauty and romance, and the Child a safe haven. Each relationship would change her ideas about herself and what she believed about love.

The Great Goddess Asteroids

Not for a moment dare we succumb to the illusion that an archetype can be finally explained and disposed of. The most we can do is dream the myth onwards.

—Carl Jung

We've already established that some asteroids are more important than others because of the size or their mythological significance. There are four asteroids in The Main Asteroid Belt that stand above all the others in significance, for both of those reasons.

The asteroids Ceres, Pallas, Vesta, and Juno rose to fame after astrologer Demetra George published her book *Asteroid Goddess* in 1986. Orbiting between Mars and Jupiter, they were discovered in the first few years of the nineteenth century and were originally designated as planets. But when it became clear there were thousands of objects in that area of our solar system there was much debate about what to do.

Should these newly discovered small objects be called planetkins? Or perhaps planerets or planetlings? Eventually, a young English Greek scholar suggested the word *asteroid*, which meant starlike. There was no precedent at the time for anything other than planets and comets, so this new classification was met with ridicule, outrage, and contempt when astronomer Herschel published his paper in the Royal Society.

The history of an asteroid's discovery and naming is considered significant in understanding its influence in the world. The abuse heaped on Herschel for introducing the word asteroid was unparalleled in the history of astronomy, an early harbinger of the response to women's quest for inclusion.

There's also a modern chapter to this story. In 2005, when Eris was discovered out beyond Pluto, the astronomical community went through another ruction. Again, it was about a new area of our solar system and the discovery of something akin to planets being discovered there. The first planet, Eris, could perhaps have been an anomaly they said. However, when, just like Ceres, it became obvious that there could be lots more of them, the planetary status quo took another hit. While the controversy in the 1800s was caused by the introduction of feminine archetypes, this latest wrangle was triggered by the introduction of dwarf planets named after indigenous creator deities.

Bringing a greater diversity to the classification system, breaking down that clear masculine hierarchy of the planets, was again claimed as an unacceptable break in tradition.

Ceres is representative of both groups. She is an archetype of the Great Mother creator deity of the neolithic era in Old Europe. When she was discovered in 1800, we saw the first wave of feminism, which focused on women's equality and the right to vote. Ceres' reclassification to dwarf planet in the early 2000s came at a time when feminism was being challenged for its lack of diversity. The time had come to recognise how closely aligned cis white women were to the established power structures. There has been a shattering of that dominant view of culture many now describe as colonisation, with previously marginalised groups now claiming the right to be part of the collective conversation.

Ceres, Pallas, Vesta, and Juno are aspects of the feminine denied a place within traditional expressions of western culture. They each hold an aspect of recovering something we need in order to heal humanity.

These four Great Goddess asteroids have been the advance forces for each wave of cultural transformation. We should never underestimate their power, in our own lives or collectively.

In the following section, I use the minor planet number that is assigned to each asteroid. The name minor planet covers everything that orbits in our solar system and is not a main planet like Mars or Jupiter. Asteroids, comets, and dwarf planets all come under the heading of *minor planet* and are all given a minor planet number to identify them. When you create a chart in Taraka you can add anything that has a minor planet number, just by adding the number to the box on the edit page.

Ceres

Minor Planet Number 1

Ceres shows what supports us to feel that life is worth living, and how we might respond if we fail to receive that nourishment. She shows how we can make sense of sudden changes in our lives when we feel a sense of loss, and what best approach we can take to find the gift and a pathway forward. She speaks very specifically to our relationship with our mothers and what it taught us about our value to others.

Ceres' Discovery

When Ceres was discovered in 1801, she was in Gate 23, Splitting Apart. With Urania in the same line, this position explains that often life sends us unexpected gifts wrapped up as loss and disappointment. Ceres is here to help us make sense of events. This is a channel of powerful insights that break apart our usual way of thinking, restructuring our reality. Both Psyche and Eros are in the same channel, Psyche in Gate 23 and Eros in Gate 43. This break, in reality, gives us the opportunity to question what we really desire at the soul level, and allow the answer to guide us through each challenge we face.

This, of course, fits perfectly with the accepted astrological meaning of Ceres as nourishment, but adds a kind of core clarity. Ceres guides us through those lonely times of loss, helping us find deep soul meaning in our experiences. Her position shows what nourishes us as we move through this difficult part of our life cycle. As we emerge, we have more clarity about what is most important, and how that new sense of purpose can be used to nourish others.

Because we are dealing with endings, Ceres advises us not to push forward while the death process is still moving through our life. We can hold to our chosen pathway, anchor into our core self, and channel the extraordinary potential available to us as we ride the chaotic part of the cycle.

In the Ceres discovery chart, the Sun is in Gate 38, Opposition, waking us up to something we initially thought was a threat, something we may have been pushing away but which we need to acknowledge. The Earth is making a channel with Pluto, focusing our attention on signs of a deeper truth that will help us overcome difficulties.

Ceres is a mystery that needs solving. She speaks to us through the opportunity to be taken by the unexpected, woken up to an ever-deepening understanding of what feeds our soul.

Ceres as the Great Mother

Much of the imbalance we experience both personally and globally arises from the removal of the mother archetype from the visible power structures of our societies. We feel its lack but can't name our loss, because we no longer have words or symbols to describe this most ancient of archetypes—the Great Mother.

Ceres supports every developmental stage of the child, but instead of moving through the cycles of life, we are encouraged to live as Pallas, free of maternal influence, born fully formed, covered in armour, and ready for battle. The very process of birth has been disrupted.

Ceres in your Design is complex, sometimes dark, and always rooted deeply in the mother-child relationship. At its best it is about nurturing, nourishment and the safety that comes from knowing you are cared for. The USA provides an example of a darker Ceres, with campaigns against socialism leaving vulnerable people homeless and everyday healthcare less accessible. In the Human Design chart for the birth of the USA, Ceres is in Gate 37 (family, food) with Nessus (a centaur representing negative family patterns) and the Child asteroid (what's needed but often not provided). The nourishment Ceres offers arises from a social fabric constructed from sharing and working to help others.

The nature of Ceres' nourishment goes beyond food, encompassing the very spark of life, as if she sits in the mitochondria of each one of us. Mitochondria are at the heart of every cell in your body, generating the energy needed to power the cell. If we don't have the nourishment, we need to feel life is worth living, why would our mitochondria generate that energy?

Ceres represents the relationship template that arises from the earliest days of the mother-child relationship and what we learn about giving and receiving, feeling safe and protected, cared for and included. How do we experience our environment? Is it nurturing? Or do we have to struggle and learn how to live as a hungry ghost?

Back in Neolithic times, the Great Mother archetype represented our participation in nature, in the cycles of birth, growth, decay, death, and regeneration (The Myth of the Goddess[8]). The swastika was originally a symbol of the four moon phases of waxing, full, waning, and dark. Ceres embodies all of these cyclical phases and guides us through each of them. Without decay and death nothing is nourished, there is no compost, nothing to feed the new. Once the male principle began to separate from the mother, the notion of death as renewal began to take on a different meaning. No longer part of the cycle of life, it instead threatened annihilation of the individual self.

Ceres reconnects us to each other and to our capacity to care for ourselves and others through the cycles of human life.

On a more everyday level, people with a significant Ceres tend to love children. I don't have a chart for Kathy Headlee, founder of Mothers Without Borders, but I'm sure she has a strong Ceres activation!

Ceres and Loss

Ceres is the Roman version of the earlier Greek Demeter—Da Mater, Mother Earth. By the time she was absorbed by the Romans, Demeter had become Ceres and taken on the much lesser role of goddess of agriculture.

You may know the story of how Ceres' daughter was stolen away by Pluto, at the instigation of Jupiter. Ceres grieved the loss of her beloved daughter, and was so filled with rage, she planned a horrible year of famine in protest. After all, she reasoned, Jupiter was nothing without humans to worship him! Eventually, a deal was brokered. Persephone would return to her mother for half of the year and spend the remainder with Pluto in the underworld.

This is a patriarchal take on the earlier neolithic seed maiden rituals, where the seed descends into the underworld each autumn to spend

---

[8] ISBN 0-14--01929201
Anne Baring and Jules Cashford, *The Myth of the Goddess: Evolution of an Image.*

time in the dark womb of Mother Earth, rising again in spring to bring the spark of life.

Ceres can indicate our ways of dealing with loss, how to manage the anger and grief, and venture into our own personal underworld to gestate anew. This was once a voluntary descent into the darkness, protected by the Great Mother. In modern times it tends to be avoided unless we are violently kidnapped by life. We often talk about this as a dark night of the soul. It has in the recent past represented a fearful time away from the father's light rather than a nourishing and necessary time in the dark womb of the mother. Persephone is the archetype of descent, Ceres of the grief and rage of enforced loss.

The Cosmic Role of Ceres

Ceres may be foundational as an ancient archetype of the Great Mother, but she also has a powerful role in modern times. In 2006, when Pluto was reclassified to dwarf planet, Ceres got an upgrade. She was much bigger than the other asteroids in The Main Asteroid Belt and met all the criteria for a dwarf planet. This was a time when astrologers really began taking notice of the asteroids. It was also a time when the nature of feminism was being defined as something that extended beyond the interests of middle-class white women.

Astronomers recently discovered that Pluto and Ceres are made of the same stuff and likely were siblings out on the edge of the solar system, till something caused Ceres to be flung into the inner realms.

There are two reasons why this is important for you and me and for understanding the rebirth of feminine leadership on this planet. This previously unknown link with Pluto has unveiled Ceres' role as a bridge between humanity and the galactic consciousness that these newly discovered outer planets is making available to us. Pluto holds open the doorway to the new class of outer planets like Eris, Sedna, Haumea, and Makemake. Ceres transmits that energy to the space where the inner individual self meets society and culture.

Pluto, Ceres, and Transformation

While we think of transformation we think of Pluto, but originally his job was more like a janitor, taking care of the underworld, making sure the dead didn't escape. Ceres worked with the living, teaching them the popular Eleusinian mysteries– the most famous secret religious rites of Ancient Greece, adapted from even more ancient agrarian rites of initiation. The Eleusinian mysteries had three parts—the descent (loss), the search (insight), and the ascent (integration). The ritual was based on the eternity of life flowing through generations, particularly through the female line.

The central theme of these rites was this:

What makes life worth living? What nourishes us? How does that idea become clearer as we go through the various cycles of life?

While Pluto has no doubt grown into his more influential role as a powerful soul transformer, Ceres sits at the border of the inner solar system teaching us how to live more soulfully as we move through the cycles of our human lives.

There's a deep and mysterious sense of the Earth's cycles in our Ceres activation. Her connection with Pluto and daughter Persephone, the Lord and Lady of the Underworld, accentuates her connection to the mystery of deep transformation through attuning to natural cycles, whether they be emotional (recovering from grief and loss), natural (growing crops, feeding ourselves and each other) or physical (maiden, mother, crone).

Rather than keeping us locked in the old cycles of keeping the gods happy, Ceres brings us into a deeper relationship with our own cycle of divine growth.

She brings us into relationship with something beyond culture and religion, the hidden dimensions of creation which we are opening to via the new dwarf planets in the Kuiper Belt beyond Pluto.

Ceres and Food

I've talked about nourishment in broad terms, but Ceres also represents our relationship with food itself. Her name forms the basis of the word cereal, and she is known for introducing grain and agriculture to humanity.

Singer, Karen Carpenter, who died of anorexia, had her Sun, Chiron, Ceres, and Hygiea in Gate 5, sometimes called Waiting for Nourishment.

Celebrity chef, Nigella Lawson, has Ceres and Saturn (career, also a tendency to tradition) in Gate 38 with Mercury (communication) in the same channel. She has minor planet Lempo, the Finnish goddess of sex, in the same gate, which says something about her reputation as a domestic sex goddess, all curves and glamour.

Ceres in your Human Design

I use Ceres in business coaching and in deep personal work. It is an archetype that urgently needs resurrecting in all areas of our lives.

On a personal level, the Design Ceres activation can show where we need nourishment but have come to accept its lack as something outside our control. I find the Personality Ceres, which operates in our conscious waking reality, gives us clues about how to reconnect with the Design Ceres.

Princess Diana had her Design Ceres in Gate 17, suggesting she needed people she could trust in her life to talk things over with. It's likely she grew up believing such people were rare. Her Personality Ceres was in Gate 24, the gate of constantly returning to her own inner compass. The nourishment she needed to make life worth living was the knowledge that she could walk away from people who were no longer good for her, relying on her own counsel to direct her when she couldn't find trusted confidantes.

On a business level, Ceres represents that special nourishing something you offer others. You will never flourish in your career if you are not drawing on that Ceres activation. We burn out when we are constantly trying to offer what doesn't also nourish ourselves.

Oprah Winfrey has her Design Ceres in Gate 47, where we can keep up outer appearances only for so long before we have to withdraw to attend to our inner world. Her Personality Ceres is in Gate 48, the gate of deep intuitive wisdom that overcomes inadequacy. Winfrey needs her personal time to tap into her own deep wisdom.

Ruth Bader Ginsberg has Ceres in Gate 37, which is part of the Channel of Family and Community. Her Design and Personality North Node (destiny) and Heracles (the hero) are also in that gate. Ginsberg was known for her championing of women's rights. Ginsberg had two devastating losses, with the early death of both her mother and sister. We can't know how these deaths shaped her, but it seems feasible that she developed the philosophy that she should make her life count. She was nourished by family and friends and found a sense of value in what she could contribute to the community.

Pallas

Minor Planet Number 2

Pallas is an ancient goddess. She is a daughter who was taken to the inner temples of patriarchy, learned all its ways and grew beyond its limitations. She represents the newly emerging global feminine leadership.

Pallas represents a kind of practical strategic brilliance that can imagine an outcome, then plan and work towards its fulfilment.

Pallas' Discovery

When she was discovered in 1802, Pallas was in Gate 46, Pushing Upwards. Jupiter is in the same channel, in Gate 29. This is, of course, wildly synchronous. In Ancient Greece Pallas was considered second in importance only to her adored father Jupiter (Zeus).

What surprised me, for such an intellectual archetype as Pallas, is how much of her discovery chart is connected to the emotional circuitry and the Solar Plexus Center. Even the channel Pallas activates is about learning from emotional experiences. The young Pallas grows beyond the linear limitations of patriarchy through the wisdom of her experience.

Pallas' Mythology

Pallas was the most important deity in Ancient Greece after Zeus. Like Apollo, she incorporated many minor deities who preceded her, giving her a wide and disparate range of skills. She was known to excel in military strategy, craft and weaving, healing, pottery, music, mediation, and law.

Her origins go at least as far back as the triple goddess of Libya with exceptional qualities of protection, wisdom, and strategy.

Her story begins with a prophecy to Zeus, that one of his children from Metis would be more powerful than him. When Metis became pregnant, Zeus decided to swallow her to prevent the child from being born. However, when Metis was giving birth to Pallas, Zeus had such dreadful pain he called for help from Hephaistos, who cleaved open his head to release the young Pallas. We never hear about what happened to Metis, but presumably she somehow escaped as well.

The story goes that Pallas was fully grown and covered in armour, ready for battle. I had always accepted this part of the myth easily until one day I read that Metis, fearing what her daughter would have to endure without her protection, created armour for her daughter while they were trapped inside Zeus.

Stolen from her mother, the goddess of oceanic wisdom Metis, Pallas was born from the head of Zeus. Her mother became insignificant, echoing the suppression of the female creatrix energy in many of ancient Greek and Roman myths. I had been working with Pallas for a few years before I even discovered her secret mother. It completely changed my take on the meaning of Pallas in your Design.

Pallas Athene

Pallas has strong connections to the famed Amazon warriors. Foster sisters Pallas and Athene were practicing their sword skills one day when Jupiter imposed his aegis (breastplate), distracting Athena who accidentally killed her sister. Athena was so grief stricken she changed her name to honour her sister, becoming known as Pallas Athene.

There is an asteroid called Athene (minor planet number 881) which is not so significant as Pallas.

The Modern Pallas

She never was a threat to Zeus, but worked in concert with him, and this is an important part of her influence in our lives. Until recently, Pallas worked in partnership with Zeus to further the aims of patriarchy. In November 2020, Pallas took part in a massive and transformative lineup—Pluto, Jupiter, and Pallas in Gate 61 meeting Uranus in Gate 24. It was a watershed moment for the archetype of Pallas, who thereafter worked from the inside to create a post-patriarchal world. I have been saying that Pallas women learned patriarchy from the inside out, and now they are using the wisdom gained that experience to create change.

We typically find Pallas in the professions of politics, law, and business strategy. I would also expect her to be significant in the

Design of a skilled craftsperson, whether it was an artistic expression like sculpture or had a design focus like architecture.

Pallas in Law and Politics

Pallas can be found in significant positions in the Designs of prominent women judges, lawyers, and politicians. In 1979 Margaret Thatcher became the first female Prime Minister of the United Kingdom. She was known as the Iron Lady, which suggests the armour worn by Pallas. Thatcher had Pallas in Gate 12, Blockage, with her Uranus in Gate 22. Her Pallas in Gate 12 Line 3 suggests an ability to look for strategy (Pallas) to push through any blockage (Gate 12) which will only backfire.

US Supreme Court justice, Amy Coney Barrett, has Pallas and Design Saturn in Gate 20, with her Uranus in Gate 57 and Neptune in Gate 34. She also has Vesta on her North Node in Gate 41. Supreme Court nominee Ketanji Brown Jackson has Pallas in Gate 55 with her North Node.

Pallas in your Human Design

My catchphrase for Pallas is "having learned the ways of patriarchy, you now apply those skills to your own higher purposes."

MacKenzie Scott, ex-wife of billionaire Jeff Bezos, has Pallas in Gate 60 (limitations) with her Design Mercury. Her Ascendant is in Gate 3 (new beginnings). In 2021 she was named the richest woman in the world. In a mirror of Pallas' relationship with Zeus, Scott was instrumental in the early days of Amazon.

Her Pallas activation, in Gate 60 Line 6, is about loosening restrictions to encourage self-reliance and confidence in others. Scott has taken a unique approach to philanthropy, giving money to organisations with demonstrated integrity and letting them decide what to do with it.

In your Design, Pallas represents the place where you have outgrown the limitations of patriarchy. It no longer gives you a space within which to express your creative brilliance. It is a place where inner reflection helps you gain real wisdom that you can use strategically to meet your goals.

Vesta

Minor Planet Number 4

I have nicknamed Vesta the "path of service" planet. She represents what is most sacred in our lives, a pathway to our greatest purpose. Your Vesta activation shows an area of great importance in your life.

She represents more of an overarching theme than concrete action. Dolly Parton has her Vesta in Gate 34, Empowerment. Oprah Winfrey's Vesta is in Gate 22, Grace. Jane Fonda's Vesta is in Gate 47, Oppression.

When I talk to people about their Vesta activation, they are often stunned by how closely it matches their attempts to express their core purpose.

Vesta's Discovery

In her discovery chart, Vesta is in Gate 46, Pushing Upward which sits on a supermassive black hole called the Super Galactic Center. This black hole carries an insatiable desire for a sacred path in life. This gate asks us to find the source of our inspiration and use it to inspire our actions.

Vesta is a virgin goddess; she has sovereignty and control over her own life. However, this doesn't mean she is all work and no pleasure! Aphrodite is also in Gate 46. This is a gate about loving our physicality, our experience of being in a body here on this beautiful planet. Vesta is the goddess living through us. Pleasure and creating through connection and relationship, are her essential qualities.

Vesta's Mythology

Vesta represents the sacred flame carefully tended by each household in ancient Rome. In Ancient Greece she was known as Hestia. I've thought a lot about that flame, what it represents, and why it was the wife's role to tend it. Rome was a patriarchal culture and most women had almost no role in public life. The home was sacred because that's where the women were.

At the heart of Vesta, we find a paradox. She is a virgin goddess, which means she belongs to herself and is not subject to a father or husband. But she is also the goddess of the hearth, home, and family. To get over this awkward splitting of the feminine, Vesta was rarely represented as human, but mostly as a flame. Demetra George refers to her as the eternal flame and this I think, gets us to something we can work with—the notion that each of us has within us an eternal flame representing our own inner hearth, home, and family.

Yet, that paradox continues to find a way into our lives. How do we live purposefully creative lives while also having the responsibilities of relationships and families? I described how Ceres holds the true transformative energy, often attributed to Pluto. We find the same with Vesta. Eros is often described as the creative spark, but he tends to act rashly and foolishly. Vesta is the true inner spark of life, a gentle sacred devotion to our own creative process.

The Vestal Virgins, known as Vestales, were held in high esteem by the Romans, but the word used for their selection was *captio* (capture). After thirty years of service, they were "allowed" to marry, which really meant that a noble marriage was arranged for them.

This idea of capture is something I see through many levels of Vesta's meaning. In times before the Roman version of Vesta, the earlier temple goddesses were the heart of any community, stewarding shared wealth, providing spiritual leadership, and representing the sacred sexual nature of the goddess. We could say that their sacred flame was captured by the Romans, who then paid the Vestales to tend it. Vestal virgins who were unchaste were buried alive. Their virginity had become a commodity and the goddess of devotion had become a servant.

When I work with people on their Vesta activation, I often find they are challenged there because of Vesta's underlying paradox. It's as if her expression requires us to break free of some agreement to serve something other than our own inner spark. The acceptable function for women has been to care for others, which in Vesta's world means keeping alive the communal flame while neglecting one's own. For men, overcoming that paradox means having permission to come home, to relax into the feminine spaces of life.

Vesta and Sexual Sovereignty

The American Medical Association reports that 42 percent of American women suffer from sexual dysfunction[9], while people spend up to 100 billion dollars each year on pornography. We are in Vesta territory here!

It's hard to imagine a time when women weren't considered property. I grew up in a time when conjugal rights were still a thing. Roe v. Wade has been overturned in the USA. We continue to live with many unexamined assumptions about our sexual expression. Sigmund Freud once speculated that civilisation is built largely on erotic energy that has been blocked, concentrated, accumulated, and redirected. I think he was onto something. Women are the only animal with an organ with no apparent function other than pleasure. I'm talking about the clitoris, in case you were wondering. Over the past few thousand years, women's sexual energy has been captured and put to the purpose of creating verifiable bloodlines. What if sex is about more than creating an heir to conserve the family fortune?

Vesta is the link between the Great Mother in her sexual form, the body of the Earth as our home, and the formation of communities based on sacred pleasure. The Vestales held a sacred function within their communities. They offered the experience of sexual pleasure. I have always seen this as providing contact with the physical body of the goddess, to be a reminder of the sacred and transformative power of pleasure. Vesta's sexual energy is sacred rather than objectified. It is not kept within the confines of monogamy but follows the sacred pathway of the goddess.

What if we were not meant to spend our days in servant-like workplaces, but live in response to our sacred inner spark? What if our sexuality was not controlled and captured, but allowed to find its own natural expression? What if we could come home to our own sacred creative selves? These are Vesta's questions to us.

Vesta is essentially a ritualised temple energy. She is no wild Dionysus, cavorting in forest orgies. Nor is she an Eros, living at

---

[9] https://www.researchgate.net/publication/13267607_Sexual_dysfunction_ in_the_United_States_Prevalence_and_predictors

the whim of her lust. This is the true containment of Vesta, she channels our wilderness along sacred pathways.

Vesta as Temple Goddess

There is a strong communal focus to Vesta. Her flame is not for the individual but for the family home. The original temples maintained and shared communal resources and upheld communal values. So, although I work with this archetype as a personal inner vocation, her energy is always destined for the greater good. People with a prominent Vesta will often work on creating beautiful, prosperous, and well-maintained homes for their own families, or others. They may be interior designers or feng shui experts, gardeners, or artists. Their purpose in life is to bring a sacred aspect to homes and communal spaces.

Vesta in your Human Design

Ruth Bader Ginsberg has Vesta in Gate 20, making a channel with her Design Mercury (communication) in Gate 34 (empowerment). She has been instrumental in arguing for women's rights throughout her time as a lawyer and jurist.

Germaine Greer, author of *The Female Eunuch*, has her Vesta in Gate 13 along with her Personality Earth in Gate 33. Greer has always championed the sexual freedom of women. She named her book for the idea that women had been sexually castrated, and that this repression cut them off from their creative power.

Sex therapist Ruth Westheimer, has her Personality Vesta in Gate 63 (innovative new trends) with her Design Sun. Her Design Vesta is in Gate 13 (fellowship, secrets) with her Design Venus. She was prolific in changing attitudes toward sex, her tag phrase was Get Some.

Vesta shows the ways in which you meet the sacred in everyone and everything, including your sexual expression. Vesta is an offering. Her purpose is to tap into the creative spark of life, the potential of feminine power to create, and to offer it up to others through experiencing the sacred nature of living in the physical realm.

Juno

Minor Planet Number 3

Juno is the spring bride who becomes the happily fulfilled wife. She is everything we long for in romantic relationships. But there are darker aspects to Juno. Not everything is as it seems in her world.

Juno is the Roman version of the earlier Greek Hera, wife of the gallivanting Zeus.

Juno's Discovery

When Juno was discovered in 1804, she was in Gate 25. The Sun, Earth, and Pluto were in the channel called Family and Community (37/40), which is about the bargains we make with our people. This is historically accurate.

When patriarchy was taking hold in Old Europe, women were bargained away as chattel-wives to keep their tribes safe from the more violent incoming people. More recently we saw Juno-type bargains where the husband was the breadwinner while the wife took care of the home and children.

The channel of Family and Community is part of the Global Cycle we've been experiencing since 1615. In 2027 we shift away from this tribal bargain which has held our families and communities in place. The influence of Juno will shift with it, and we can already feel the quaking of change in the relationships between women and men.

Hilary Barrett's I Ching[10] says this about the Sun's position in Juno's discovery chart:

"When you let go of your requirements for how things (especially relationships) must be, you release the situation's latent potential."

Juno is in a gate that supports us to disentangle from negative influences. How do we maintain our innocence, despite life's shocks and initiations? Should we withdraw into a relationship where we will be taken care of? Perhaps it's the relationship itself that creates the shocks?

---

[10] https://www.onlineclarity.co.uk/about/

In astrology Juno represents marriage and relationships of all kinds. Ideally, these relationships are built on mutual respect, equality, and loyalty. The shadow side of Juno is obvious when we look at the mythology of Juno/Hera and Zeus. The mutual bargain was made, the marriage took place. But Zeus didn't keep to that bargain and Juno found she had no sway over him. Almost every story about Juno and Hera arises from Zeus' infidelity and her attempts to gain revenge. So much for the blushing June bride!

When Zeus dallies with another woman (often a nymph, Zeus was really the original nymphomaniac!) Hera/Juno takes her revenge on the other woman. It's almost as if Juno is showing us what kind of bargains we make and how we look the other way when they are broken, blaming those outside the relationship because we don't want to address the core issues.

Juno's Mythology

The name Hera comes from He-Ra, the Earth, and there is some connection between Hera and Gaia (Anne Baring[11]) She was the goddess queen who chose her fresh new king each year. In Ancient Greece, the word hero meant one who has gone to the goddess and sacrificed to his queen. (Barbara G. Walker).

Originally the goddess chose the god, but mostly we see Juno coming second to Zeus. If we go back further, we find the original sacred marriage between Heaven and Earth. That elemental ritual was shattered in Ancient Greece. Aphrodite was handed love and romance. Hera was left with the formality of marriage, but little else. Jane Harrison said Hera had been forcibly married but was never really a wife (Ann Baring). She could not even claim to be a mother. Despite having children, Juno's focus was always primarily on her husband and the breach of their agreement.

Wrapped up within Juno is the whole notion of monogamy. But it turns out that the monogamous long-term relationship is an outlier, a cultural fantasy. Recent research makes it clear that adultery is

---

[11] ISBN 0-14--01929201
Anne Baring and Jules Cashford, *The Myth of the Goddess: Evolution of an Image.*

common in every monogamous human society ever studied (Ryan[12]). It seems that Juno opens up the question of what is a realistic mutually satisfying sexual relationship if we strip away the ancient restrictions of ownership and control of women!

The mythological nature of Juno can be found in her role as the queen of immortality and sacrifice. She was the Queen-Bride in the sacred marriage who took a sacrificial bridegroom in spring to renew her virginity and ensure food for the coming year. The old goddesses were the matrix of the life cycle—birth, life, death, and regeneration. But Zeus seized control of the creative matrix held by the goddess, and the nature of women narrowed to daughter, wife, mother. Juno is the wife.

Juno is usually seen as an indicator of what your perfect relationship might look like, and what is blocking you from having it. As the Global Cycle shifts, it's likely that the thing stopping you from having that perfect relationship is the cultural assumptions about marriage, monogamy, ownership, and control.

Juno in your Human Design

I work with Juno as an indicator of where you should pay attention to your need for mutual respect in your relationships, both romantic and business. This is where you may accept bargains that work against your better interests, perhaps motivated by unconscious needs for security or status. Once you've accepted the bargain, you get stuck because of a misplaced loyalty that is often not reciprocated.

Maria Shriver has Juno in Gate 34 with her Personality Venus; her Mars is in Gate 57. It's likely she looks for experiences of instinctive power in her relationships. Her Zeus is in Gate 29 with her Pluto. It's no surprise that she married Arnold Schwarzenegger! Speaking of which, his Juno is in Gate 10, making a channel with his Nodes in Gate 34 and 20, which are about personal empowerment. We can never know what goes on in the relationships of others, but this seems to be a very Juno relationship! From all accounts they had a long and happy marriage,

---

[12] ISBN 978-1-921844-24-9
Christopher Ryan and Cacilda Jethá, *Sex at Dawn: The Prehistoric Origins of Modern Sexuality.*

representing the very best of Juno. But then, adultery happened. Schwarzenegger fathered a child with their nanny. Shriver divorced him.

The Asteroids

So far, you've been exploring the goddess asteroids made famous by astrologer Demetra George—Ceres, Vesta, Pallas, and Juno. It's time now to explore beyond those four powerful archetypes, to wander into the realms of your inner world and explore your Persephone, Europa, Isis, and more.

The asteroids in this section may not be as generally influential as the Great Goddess asteroids, but their position in your Human Design chart can make them more important in your life. For example, if you have Ceres in an empty channel in an undefined Center, she may be dormant and have little influence in your day-to-day life. On the other hand, you may have Artemis in the same gate and line as your Moon, making this moon goddess a constant companion. You'll find you can be quite a loner, prefer animals (dogs in particular) to people and love being in nature.

Every asteroid can be significant at times. If the Sun makes a channel with that unconnected Ceres, you may find yourself confused by a sudden rush of grief, or a desire to eat more than usual. These times are important. Activation of these more dormant asteroids can bring up old memories from previous times of activation, or just give you a wider range of experiences. An activation to your Artemis, however, will feel personal and connect with your daily lived experience of self.

It can be overwhelming to start reading about the asteroids. It's tempting to try to take them all in at once. That can be fun, but as you move more deeply into exploring the impact of the asteroids on your life, it's best to use the process I've laid out in Chapter 3.

Aphrodite

Minor Planet Number 1388

Known as the Goddess of Love, Aphrodite was born out of conflict between her mother Gaia and father Uranus. She knows how to create harmony from friction and the tension of opposites.

In the Book of Genesis, a failure of obedience by Adam and Eve brought down the punishment of painful toil for both. How different is Aphrodite's view of the world?

> *She crossed the pebbled beach and wandered over hills and plains, seeking out all living creatures. Magically she touched them with desire and sent them off in joyful pairs. Everywhere, Aphrodite drew forth the hidden promise of life.*
>
> *—Spretnak, Pagan Meditations*[13]

The desire of Aphrodite draws forth the tension that arises from the other, from what is different. She reminds us of a time when we weren't threatened by what is different, but intrigued, seduced. Aphrodite fills us with desire to join with the difference, to learn something from it so we can grow beyond our own small, contained world.

In our Human Design, much of what we attribute to the planet Venus actually resides in our Aphrodite activation. Venus, originally a minor deity of fields and gardens, came to be associated with the ancient and more powerful Aphrodite. The planet Venus is more significant in our Design, but the asteroid Aphrodite shows how we connect with our true sensual nature, the vital tension that creates new life.

Aphrodite's Discovery

Aphrodite was discovered in Gate 3 Line 3. Gate 3 is the result of the yang in Gate 1 joining with the yin in Gate 2 and represents all the tumultuous chaos arising from their coupling. The gate is often

---

[13]https://www.charlenespretnak.com/lost_goddesses_of_early_greece___a_collection_of_pre_hellenic_myths_117206.htm

called Difficulty at the Beginning which Richard Wilhelm describes as a teeming chaotic profusion, like a thunderstorm. For anyone who has been in love, that description fits perfectly. Lost in the wild country of our own desires, the Line 3 suggests we slow down and allow ourselves to be guided by our sensual nature.

Eros, the son of Aphrodite, tells us what can happen if we let our lust run away from us. There's a kind of numerological perfection in finding that in Aphrodite's discovery chart, Eros is in Gate 13.3. All those 1s and 3s! The Eros activation reminds us to take time to reconnect with our original intention when we feel desire, not to feel forced to abandon or compromise ourselves, but to find ways to create harmony.

Aphrodite's Mythology

A most ancient goddess, Aphrodite danced her way to Greece from Mesopotamia via Crete. The Greeks converted her immense creative power, turning her into a beautiful sex object. It would be a mistake to underestimate her in this way.

Consider what happens when we give up on our own desires and allow the desires of others to shape our world.

Aphrodite brings seduction, a longing for intimate touch, and a desire for pleasure and play. She shows us that duality is not the endpoint, but the beginning—a reunion of opposites. The polarity of light and dark makes no sense when we are watching the sun rise and set. Hot and cold are not two separate things but join to give warmth.

In times of change, Aphrodite tells us not to rush forward, but to reconnect with what we find beautiful, seductive, and lovely. When we experience difficulty, she suggests we look for a new way of living that is more informed by pleasure than discipline.

Aphrodite's world arises from desire. As with the mystical teachings of the Kabbalah, Aphrodite teaches us that our desires are a direct message from the goddess, communicating what is important to each one of us. Desire guides us, not always in easy ways, to find who we want to become.

In relationship, Aphrodite will not bow to patriarchal rules about control of women's sexuality. She unapologetically takes her pleasure where she chooses. She enjoys romance but is not the marrying kind. There is nothing skimpy about Aphrodite, she is about the full-bodied enjoyment of life.

We have been convinced that sexuality is such a disruptive force, but Aphrodite is a sophisticated and civilising influence. It was typical of the ancient Greek and Roman treatment of the ancient goddess to paint female sexuality as a chaotic disturbing force. But Aphrodite's beauty is a different kind of arrangement to the punishing control of the patriarchal god. To the ancients it was the briny sea from which everything arose, and Aphrodite teaches us that this way of living has its own natural arrangement.

Aphrodite in your Human Design

Beyoncé has Aphrodite in Gate 5, with her Design Earth. In Line 3, that Earth activation sits right on the Great Attractor, which tends to draw great attention. She has Design Venus in the same channel. All that sensual attractive energy in the Channel of Rhythm is helped along by Jupiter (good fortune) in the Channel of Dancing (Gate 48). She also has the muse of the dance, Terpsichore, in a channel with her Personality Venus!

Aphrodite shows where you need to let your sensual nature out to play, where your desire is rearranging your life away from hard work and into a deeper level of creativity that embraces the feminine ocean of desire and possibility. She reminds you to be informed by your own desire, not shaped by the desires of others.

Apollo

Minor Planet Number 1862

We might marvel at the attributes of the sun god Apollo, what does he not excel at? Archery, music, dance, prophecy, healing, he is the patron of seafarers and protector of refugees, giver of laws, inventor of the lyre, he was the chorus leader of the musical Muses and was known as the averter of evil, particularly plagues.

Like Pallas Athene, the wide scope of Apollo's accomplishments reflects the absorption by Ancient Greece of many earlier cultural practices and myths. But what does he represent in our Design?

Apollo's Discovery

When Apollo was discovered, he was in Gate 50, making a channel with the Sun in Gate 27, confirming his association with the Sun. Artemis, his moon goddess twin, is in Gate 15, exactly opposite the Moon in Gate 10.

Gate 50 is where we come face to face with our values. Apollo represents the creation of culture. This is the point where we make a choice to express our highest values, to embrace our whole self, to face the darkness and acknowledge its power or excise it completely from conscious awareness.

Apollo mediates the power of the Sun, ensuring we recognise our real nourishment as both light and dark aspects. He has something to say about HOW we mediate that space between humanity and divine order. We can't do it by denying our nature, for example by pretending we don't love who we love or trying to be who we are not. There is something essentially good in Apollo, when he turns up it's as if the sun shines on everything around him, but we also have to notice the shadow he casts.

There is a Uranian object called Apollon, also named for the god Apollo, which has a similar meaning to the asteroid Apollo.

Apollo's Mythology

Apollo's many accomplishments probably arise from earlier gods like Paean, a shaman who used music to cure disease, and early agricultural deities who hunted field rats, explaining his association with plagues. His name may be connected to the initiation of young men at festivals called *apellai*. Early versions of his twin sister Artemis had a male helper called Master of Animals, another role which Apollo may have taken on.

In all of this absorption of earlier cultures, Apollo stands out as somewhat lacking in his own personality. He was never the sun, but the vessel conveying it across the heavens. He was never the king, but always the king's favourite son.

This is key to how we work with this asteroid. Apollo is the young man, experimenting with life, discovering what he loves, where he excels, and harmonising those things into something new at the highest level.

Always associated with light and the sun, Apollo was instrumental in slaying the Python (representing the goddess religion) to gain control of the Delphic Oracle. Apollo demanded of Zeus that all forms of divination be made inferior to his, a wish which Zeus granted him readily. Hekate's dark dreams and visions were cast aside as inferior. All oracular voices were thereafter filtered through the light-filled voice of Apollo.

So Apollo can represent the dark/light split, giving the light authority over the dark. But there is more here, to do with gender and desire.

Both Apollo and his sister Artemis were gender-fluid. Although Apollo was said to have fathered children with each of the Muses, his great love was Admetos. His effeminate behaviour with Admetos was said to have embarrassed his family. Apollo was also lover to Adonis and Hyacinth. Many gay people know what it means to act straight to please their families, and I wonder if Apollo has something to add there. Apollo can play into the idea of staying in the light, keeping close to the parental approval, not being captured by the so-called darker desires.

Apollo in your Human Design

Prince Harry has his Apollon in Gate 48, and the asteroid Apollo with his Design North Node (destiny) in Gate 16. A very Apollo man! This is the channel for celebrating the rituals and skills that build our cultures.

I was interested in comparing Harry with Prince William, who at first glance seems much less skilled and more submissive to the influence of the Crown (Zeus). Prince William has his Personality Apollo in Gate 20, connecting his Uranus, Saturn, and Earth to the Throat Center. This suggests he is suppressing a whole lot of personal expression, but the potential is there for him to be a profoundly gifted leader if he can balance his Uranus (freedom, revolution) and Saturn (stability, tradition).

With both Prince William and Prince Harry there is a darker side, the colonial history upon which their royal lineage is built. It seems they both have a part to play in the changing culture that arises from that emerging recognition.

On a personal level, your Apollo activation shows where you have direct access to the mind and will of the divine, and the ways in which you can best share that with humanity. As a cultural story, Apollo shows a kind of stunted leadership that is always looking up to Zeus, trying to stay close to power. Apollo's choice is to act instead from his own personal higher values.

Ariadne

Minor Planet Number 43

Ariadne is inextricably linked to the Minotaur, a bull-like monster dwelling at the center of the labyrinth. I find people with a prominent Ariadne activation are very drawn to labyrinths and have often built one on their own land. Ariadne is best known for showing Theseus how to slay the Minotaur and escape the labyrinth by following what's known as Ariadne's Thread.

Ariadne's Discovery

The asteroid Ariadne is in The Main Asteroid Belt and was discovered in 1857 when it was in Gate 50. Called The Cauldron, this gate describes an alchemical ritual for receiving guidance from the ancestors and spirit realm at times of great change.

Sisyphus is also in Gate 50. This speaks profoundly to Ariadne's process for change. The asteroid Sisyphus is about the specific ways in which we will tend to resist change. Compare Sisyphus, cycling endlessly on his wheel, with the journey of Theseus as he heads into the labyrinth with Ariadne's guiding thread. With Ariadne's help, we can walk towards our own demons, represented by the Minotaur, face the trapped darkness, transform it, and then return to our everyday world. The labyrinth represents the sacred center where deep transformation can happen and Ariadne's thread guides us back into the world with this new aspect of self fully integrated.

Ariadne's Mythology

Ariadne was a princess of Minos, and part of a multi-generational story that centers around sacrifice and bulls. There was an earlier time when an annual sacrifice of the queen's male consort assured a bountiful harvest. Later, the human sacrifice was replaced by a sacred white bull, representing the power of the fertilising masculine energy entering into the womb of the mother.

But as time went on, even the nature of the bull changed from its original sacred and sacrificial nature, becoming instead a terrifying monster, imprisoned at the heart of the labyrinth.

The famous Chartres labyrinth is thought to have had an engraving of the Minotaur at its center. It had become Christianity's version of the devil. Walking the labyrinth was a contemplative ritual, with Ariadne's thread being the pathway through the labyrinth. The process of walking represented a choice to follow Christ so as to avoid the devil.

The Minotaur was not the only monster Ariadne is associated with. In one version of her story, Ariadne was killed by Perseus who turned her to stone using the head of the monster Medusa.

At the beginning of the story, when Ariadne was a young woman, her father put her in charge of the labyrinth. She spied Theseus amongst the sacrificial victims about to enter the labyrinth and fell instantly in love. He had come to try and kill the Minotaur, and Ariadne secretly gave him a sword and thread to help retrace his way out of the labyrinth. When he succeeded, she eloped with him.

The transformation of the sacred becoming monstrous is a familiar one that we see with many of these archetypes. The most obvious is Medusa, the beautiful young maiden who becomes a stone-cold monster. We have become conditioned by our culture to follow only the light and to avoid the darkness, which is now characterised as hell. Ariadne gives us access to the original core space of transformation. Like the seed maiden Persephone, we can voluntarily venture into our own dark depths and find a way to approach the fertilising power hidden there. Without a ritual process for making a connection, life becomes a kind of maze. We lose Ariadne's thread. We externalise the monster and begin to fear it so deeply it can turn us to stone à la Medusa.

The loss of a ritualised and sacred process leaves us struggling to find our way. The words sacred and sacrificial have the same root, and here they represent what we need to sacrifice in order to stay on a sacred pathway.

Imagine standing at the entrance to a labyrinth, contemplating an issue before you begin to walk slowly towards its center. You set an intention to reveal something that has been hidden, and as you follow the pathway, Ariadne's thread, you surrender what has been preventing you from having clarity. You want to see your monster, to take what has been turned to stone out of fear and bring it back to life. On the journey out of the labyrinth we take in new ideas and perspectives as the darkness and demons are integrated. Labyrinths are all about wholeness.

Ariadne in your Human Design

Albert Einstein had his Ariadne in Gate 12 (blockages) with his Design Earth. His Design Sun is exactly aligned with the Galactic Center in Gate 11.5. When things are darkest, we can turn and walk back towards the light, and blockages disappear.

Britney Spears has Ariadne in Gate 32 (transformation) with Sisyphus, Hekate, and Pluto. In Line 1, Ariadne is telling Spears to steady herself by following her own course, what feels most natural to her.

In your Design, Ariadne represents the intentional alchemical practice of surrendering to your own demons, the natural sacrificial/sacred pattern of your unique life. Ariadnes is like the in and out-breath. As we follow her thread, we create a life that fits us, rather than forcing ourselves to fit into a life that may be defined by religious beliefs.

Ariadne wants us to walk on experimental ground, to explore life through walking our inner sacred space, to see it reflected in the pattern and process of the labyrinth. Through our attention and intention, we can choose what goes into our own alchemical container and surrender to the release of what we no longer need.

Artemis

Minor Planet Number 105

Artemis prefers wild forests, clear streams, and untamed creatures to the luxury of a home. She is the young lunar goddess, huntress, and counterpart to the sun god Apollo, who is her twin. She rebels against any containment or social obligations, often preferring the company of animals to people.

Artemis represents the particular time of girlhood when we can venture away from our family but have not yet fallen in love. She is a virgin goddess, where the word virgin means the sovereignty of self. I find Artemis also often returns to the lives of women once their children have left home, a time when they can release domestic obligations and begin again to experiment with expressing their wild independent self.

Artemis' Discovery

Discovered when she was in Gate 17, there is a story here of feeling unsure about whether we really want to grow up. We might prefer to stay in the unformed potential, the mental construct of the Ajna Center, rather than risk who we could become.

Both Artemis and Apollo face this dilemma, indulged and protected by their father Zeus, there is a kind of insulation from consequences or responsibility. We can see this with Ivanka Trump, who has Artemis in a channel with her Design Sun (the father). I've noticed this sense of feeling lucky, blessed, even entitled, when I work with both Artemis and Apollo. There can be a tendency to look to a father figure to provide support, encouragement, and finances. There's nothing wrong with that unless it prevents us from growing up. By looking at your activations for both Jupiter and Zeus you can explore ways to integrate that father energy into your own psyche.

The positive attribute of Artemis is her passion for stepping outside of the accepted rules of society and testing herself in the real world. A healthy Artemis has the fortitude to stay true to herself because she can trust she will be supported in that exploration of self.

In her discovery chart, the Moon in 46 is making a channel with the North Node in 29, an energy that moves through difficulties by knowing what to commit to. Saying no is a gift that Artemis has, she has no desire to be caught up caring for others but is a good ally in difficult circumstances. Jupiter is in Gate 21 and Zeus in Gate 45, so we have the father figure in the channel that represents the community leader, the head of the household. The challenge for Artemis in integrating this energy into her own psyche is not to give up her unique and unusual nature to her father's ideas, but to find her own pathway, by following her North Node destiny of knowing what to commit to (Gate 29).

Artemis' Mythology

As the huntress, Artemis knows what she is aiming for, and she shoots straight. Greta Thunberg, a Swedish environmental activist, has both her Artemis activations (Design and Personality) in channels with her Moon.

Artemis has a love for nature, particularly forests and streams. She is often found accompanied by her dogs. I remember doing a reading for a woman who had Artemis in the same gate as her Moon. I asked if she had dogs and was surprised when she said she had never wanted to have dogs although she loved them. Then she paused in thought and told me she had slept with a giant painting of a dog over her bed her whole adult life.

Artemis is happy alone, but also likes to hang with her crew, most often a group of girls. When I say girls, that crew could be the girls you went to school with, even though you are all now in your forties. Romance is not so important for Artemis, and she is not likely to abandon her peers for a relationship.

Artemis is an androgynous archetype. We see this in the context of twin Apollo (blending masculine and feminine) but also in her refusal to wear dresses or engage in other culturally approved feminine ways. The twin energy also shows up in the paradoxical nature of Artemis, both hunter and hunted.

Artemis in your Human Design

Author Virginia Woolf had Artemis in Gate 43 with her Design Mercury. I'm not surprised by this, she had a restless need to wander freely in her own perspective, hunting down a truth that made sense to her. Her husband Leonard Woolf was her Zeus, taking care of all the practical matters that allowed Virginia to roam with little restriction. Virginia's Zeus activation was in Gate 15, which is Leonard's South Node (fate, karma).

Virginia Woolf had a sexual relationship with Vita Sackville-West, which also ties in with the gender-fluid nature of Artemis.

In your Design, Artemis represents a young aspect of self who is private, independent, feisty, and confident. She isn't interested in love, in being overly sociable, or meeting anyone's expectations.

The paradoxical nature of Artemis may be confusing, she is both protector and hunter, midwife and slayer of young women, wild and yet accomplished at getting what she needs from others. And while she is very independent, she has a close relationship with her father, who understands and supports her unusual lifestyle.

Her position in your Design shows where you may find relationships too restrictive, where you need to allow yourself time and space alone, or with your crew.

Astraea

Minor Planet Number 5

Astraea was the Greek Goddess of Justice, Innocence, Purity, and Precision. Her name means star-maiden. She represents a spirit of renewal, particularly of culture, brought about through an uplifting of human morals and values.

Astraea's Discovery

When she was discovered in 1845, Astraea was in Gate 16, Enthusiasm, with Vesta (path of service), Persephone (mastering the dark night of the soul), Minerva (pleasing the father), Nessus (negative family patterning) and Transpluto (perfectionism). They connect with Epimetheus (hindsight) in the Gate of Depth.

Quite a gathering!

Let's explore what all these archetypes might be doing bunched up together in this channel. Persephone and Nessus represent sudden shocking events that force us to find new ways to define who we are. With the wisdom (Gate 48) of hindsight (Epimetheus), we can see what might have led us to be caught up in such difficulty. The key is in listening to our own quiet deep wisdom (Gate 48) rather than lending our enthusiasm (Gate 16) to the goals of others (Minerva and Vesta) and getting caught up in overwhelming, even shocking, situations. The Moon is in the Gate of Shock.

Astraea's Mythology

Despite her sensitivity, Astraea was the last of the immortals to remain at the end of the Golden Age. Eventually even she had to leave, as the increasing wickedness and blood-stained darkness of the Iron Age sent her scurrying for the safety of the stars, where she became the constellation Virgo. According to mythology, Astraea's return to Earth will herald the dawn of a new utopian Golden Age, bringing an end to human suffering.

Astraea in your Human Design

Comedian and actor Jim Carrey strikes me as a very sensitive person. His Astraea is in Gate 39 along with Orcus (living our soul's purpose) with his Chiron in Gate 55 making up the channel. As I write this, Carrey has announced his retirement from acting. In *Variety Magazine* he is quoted as saying "I really like my quiet life and I really like putting paint on canvas and I really love my spiritual life and I feel like—and this is something you might never hear another celebrity say as long as time exists—I have enough. I've done enough. I am enough."[14]

In your Design, Astraea shows a tendency to stay too long and a difficulty in letting go of things that are no longer good for you—relationships, jobs, homes. There is a false enthusiasm brought on by a deep need to be of service. But this enthusiasm is misplaced. We can see from the example of Jim Carrey that Astraea also brings the ability to only choose what's most important.

Astraea is a catalyst for a new level of sensitivity. A prominent Astraea could be an indicator of being a Highly Sensitive Person (HSP). You need to recognise and value this gift. If you stay too long in circumstances that overwhelm you, it's likely you will end up in trauma and shock. You are here to bring refinement and purity to those situations where people are open to being uplifted.

Astraea calls for a recognition of the gift this sensitivity offers. Astraea is a powerful archetype for uplifting human morals and values, but don't allow your enthusiasm to override your sensitivity.

---

[14] https://variety.com/2022/film/news/jim-carrey-retiring-acting-1235220945/

Child

Minor Planet Number 4580

The asteroid Child shows where you are connected most strongly to the experiences of being a child, both the traumas and the joys. This activation in your Human Design is a place where you can explore reuniting with your inner child in a way that brings her magic into the present.

Child's Discovery

The asteroid Child was discovered in 1989 when it was in Gate 32, making a channel with the asteroid Echo. Echo repeats herself over and over, but no one is listening. This channel of Transformation speaks so strongly of the desire of that child to be recognised for her contribution to improving the lives of those she loves.

There is a trauma of staying stuck in what we didn't get–recognition, support, love. Our child is trapped for lack of recognition of her capacity to see what is most needed in our day-to-day lives for ongoing transformation.

The Moon, Diana, and Hekate are in Gate 21, which describes tough long-standing issues. We should persist in working through these issues, not allow the difficulties to drive us away. There is good reason for our recovery work, and a worthwhile outcome in the end.

Child's Mythology

There is no mythology for this asteroid as it doesn't come from the Ancient Greeks or Romans. It was named after Jack Child, a software engineer and Asteroid Project Manager at the World Space Foundation, as a mark of respect and appreciation for his many years of introducing newbies to the work of Asteroids in Space. Nothing to do with mythology, and not even anything to do with children although it definitely has a connection with the idea of caring for those who are younger and more inexperienced.

Child in your Human Design

Greta Thunberg has the asteroid Child in Gate 38 with her Personality Sun. She is often dismissed as a child despite her powerful transformational work on the global stage.

Michael Jordan has a children's charity offering big brother and sister mentoring. His Child asteroid and Pluto are in the Channel of Family and Community (37/40).

Shakira has a children's charity focusing on educating children in her native Columbia. Her Child asteroid and her Earth are in the Channel of The Beat (2/14) which is about encouraging others to develop their skills so they can change their direction in life.

I use this asteroid in every reading I do, as I find it shows a core childhood wound that is probably a part of every issue you face as an adult. There is a real sense of pain and child-like passivity here and you may find it takes some time and gentle probing to discover how to bring your active adult self in to support this young part of you.

In its healed state, this activation shows what we need to be healthy in our child-like nature, and how we can weave it into our daily lives. It is often a place where we feel vulnerable and in need of support and can eventually be highly creative and connected.

Circe

Minor Planet Number 34

Circe is a magician, known for her vast knowledge of potions and herbs. She has a strong desire to help others, but sometimes her motives can be self-serving. When our intentions are clear, Circe brings a strong capacity to facilitate the good of others, through magic, energy work, herbs, and all kinds of alchemy.

Circe's Discovery

When the asteroid Circe was discovered, she was in Gate 32 Line 6, a gate about making progress by working on what we already have in our life rather than trying to magic ourselves into a completely new life. Not even a magical island of feasting and wonderful sex! We create better outcomes by staying in touch with reality, patiently working on transforming the life we have, and noticing each day what can be transformed. This is the magic of Circe—an active creative process grounded in patient reality.

Circe's Mythology

Circe was the daughter of Helios, a seductive and treacherous goddess with beautiful hair and flashing golden eyes that shot out rays of light. She was highly skilled in magic, particularly in the use of herbs, and enjoyed taking revenge against those who displeased her. She was well known for transforming men into beasts.

Circe is best known for luring Odysseus and his sailors to her island and serving up a feast. She offered the sailors wine from an enchanted cup and turned them all into swine. They stayed on Circe's island feasting and drinking wine for one year. Circe eventually released them and gave them instructions about how to return home.

In fact, Circe herself goes through many transformations. She was originally a very minor goddess but was written about repeatedly until she had become a predatory woman, a monster who took away people's minds, and a wicked witch. In all this, we can see the shift in consciousness from a wise woman of herbs and potions to the fearful bearer of dark magic.

Cersei Lannister in Game of Thrones is said to have been based on the archetype of Circe, the evil queen with the power to reduce men to their baser instincts.

Circe in your Human Design

The word bewitching comes to mind when I think of Circe. Beyoncé has Circe on her Design Mars in Gate 8. However, we shouldn't objectify Circe by thinking of her as a simple temptress. Beyoncé has her Design Circe in a gate with her Design Mercury, which suggests she knows how to communicate Circe's magic. Marilyn Monroe has her Circe in a channel with her Pallas, suggesting a clever strategic use of her bewitching qualities. Oprah Winfrey has Circe in Gate 37 (creating community) with Europa (the Empress), making a channel with Pallas (creative brilliance) and Medusa (overcoming fear) in Gate 40 (finding solutions). This suggests a business-alchemy aimed at supporting women to find a better place in their communities.

Your Circe activation shows where we can magic away our enemies or those things that displease us. But we can't put to sleep those troublesome parts of ourselves or live in perpetual chaos while hoping things will magically transform. Circe is a place where we have incredible skills of transformation, and where we must become conscious of our motivation and intention.

Eros

Minor Planet Number 433

Eros represents true desire. We live in a culture that monetises that desire, persuading us that we must buy the next shiny thing in order to feel fulfilled. Eros guides us away from that false desire and the effort to fulfill it, and towards real pleasure.

Eros' Discovery

When it was discovered in 1898, Eros was in Gate 49, Revolution, with the Earth (Gaia in the creation myth). Revolution changes the fundamental building blocks of reality, bringing new worlds into being. The minor planet Chaos was in Gate 41, Decrease, which is the only start codon in the Human Design chart. I do love these synchronicities in the minor planet discovery charts!

In Line 4 of the Gate of Revolution, Eros represents a radical change. If that change comes from the highest of motives all will go well. If we are driven by ego and personal gain, it's likely to go horribly wrong. It speaks to us of the difference between a lower-level lust and a higher-level creative desire.

In the discovery chart for Eros, the Sun is in the Gate of Youthful Folly (4), the inexperienced fool. There is nothing wrong with inexperience as long as we are open to learning along the way. Relationships are usually our biggest teachers, and sexual relationships most particularly.

The discovery chart speaks of the power of Eros in our lives. If we work consciously with this energy, it will challenge our ego-based worldview, tapping us into a whole new way of working with desire and passion.

Eros' Mythology

Although there is only one asteroid called Eros there are two versions of Eros in mythology. The original version was one of the first primordial beings, along with Chaos, Gaia, and Tartarus (the Abyss). One of the fundamental causes of creation, Eros mated with Chaos and brought forth the human race.

The second version of Eros is closer to our idea of Cupid. Subject to the whims of his powerful mother Aphrodite, this version was a mischievous young man who fell in love with the beautiful Psyche.

Together these two aspects of Eros form an archetype of passionate sexual love as a creative force. Eros energy can evoke the trickster, drawing us irresistibly to people, places, and experiences we should sensibly avoid, but simply cannot. At his highest frequency, he gives us desire for those experiences that encapsulate our true purpose.

Eros in your Human Design

While we might expect Eros to be all about relationships, his scope goes beyond that, to the act of creation itself. When we have lost our passion, feel something is over and can't seem to find our way, it's as if we are with Chaos awaiting Eros' influx of new desire. Eros represents sexual passion, even obsession, but also the passion of any kind of creative enterprise.

Feminising academic Germaine Greer, author of *The Female Eunuch*, has Eros (passion) and Vesta (what you are devoted to) in Gate 13 (fellowship, secrets, and allies). Her book said all the things women were thinking but not saying out loud (Gate 13—secrets). She was also very sexually open (Eros—erotic energy; Vesta—sexual sovereignty). Greer was often in trouble for her passionate outspokenness.

Nelson Mandela has Eros in Gate 15 (humility and integrity) with his Jupiter and Pluto. His passion was to create deep change (Pluto) to the unconscious beliefs (Jupiter) that required a commitment to a long period of self-development. Gate 15 can move very slowly, and no doubt Mandela's twenty-seven years in prison were instrumental in his having the humility and integrity to bring about the powerful change he desired.

We also find the asteroid Hera (wife) in that gate, suggesting his desire (Eros) was kept within marriage, or perhaps that he was guided by a consciousness of the feminine.

In your Human Design, Eros represents an openness to learning from the experiences, good and bad, that arise from passionate desire, whether they be sexual or not. This is the place where you can be driven by your erotic impulses and need to choose carefully for creation rather than destruction. It's where we can be in love with the idea of love, wanting to draw forth life from the void of Chaos.

Europa

Minor Planet Number 52

Europa was so very beautiful that Zeus found her irresistible. He transformed himself into a white bull and persuaded her to ride on his back, then whisked her off to Crete where he made her queen. Once there, she created a stable and prosperous kingdom.

Europa's Discovery

Europa was in Gate 40 when she was discovered in 1858, a gate that offers solutions if we can look beyond the belief systems we've been raised within.

Europa's discovery chart encourages us to move with situations without getting caught up trying to make things right. We could imagine Europa as being overpowered and tricked by Zeus. But we have to account for how life turns up for us, opening new doorways. Asteroid Lachesis is in the same channel as Europa. Lachesis is one of the Fates, interrupting life so that it heads off in a new direction.

Europa's Mythology

The motif of the bull, a symbol of the ancient and all-powerful Earth goddess, is deeply woven through Europa's mythology. Europa is much more ancient than the Greek myth and was likely related to Demeter and the ancient moon goddesses like Astarte.

Her most important myth is of being carried off by Zeus. Kidnap and rape are two of the most common themes in Greek and Roman mythology, but it seems in this case that Europa was quite happy with the seduction and was determined to make the very best of her new circumstances.

Europa seems to me to have the same energy as the Empress in the tarot, steadily creating a calm and prosperous empire.

Europa in your Human Design

Ghislaine Maxwell has Europa in a channel with her Design Moon, which perhaps created a predisposition to be easily seduced and to seek stability (Moon) in wealthy, powerful men (her father, Robert Maxwell, and her business partner Jeffrey Epstein). She has recently been convicted of trafficking young women (Moon) for Epstein (the seductive Zeus disguised as a harmless bull).

Oprah Winfrey provides another view of Europa, which makes a channel with her Design Earth. For Oprah, the asteroid Europa has given her a personal pathway to construct and direct the course of her own life, building stability and financial security.

In your Design, Europa shows where you have the ability to make the most of any situation to create a calm and prosperous life. Rather than being a victim, you should take responsibility for circumstances so as to make the most of them.

Hebe

Minor Planet Number 6

When people turn up at their Human Design session asking how they can best be of service, it's their Hebe asking the question. The asteroid Hebe is strongly associated with being useful and helping others. One of her key roles was to serve a delicious ambrosia of eternal youth to the immortals.

Hebe's Discovery

The asteroid Hebe was in Gate 26 when she was discovered in 1847. If we think of Hebe as our key to eternal youth, perhaps we could heed her message here—that we should not lightly fritter away our resources and vitality. If we have the self-control to overcome our desire for short-term outcomes, we can accumulate the wisdom to eventually achieve greater things.

The Sun activation echoes this idea. In the Gate of Stillness, our development over time (Sun) comes from staying true to our unique nature even during difficulties. This is how we can nourish ourselves and not waste our resources on activities that won't serve us in the long run.

Hebe's Mythology

Hebe is the goddess of youth who ensured the immortality of the gods by serving them ambrosia to keep them eternally young. We shouldn't confuse Hebe with an everyday servant, she was much more than that, the source of vital life-force and immortality.

Hebe in your Human Design

When I was thinking of people who might have a prominent Hebe in their Design my mind went to Rupert Murdoch, who seems to be ageless. His Hebe sits in the complementary gate to his Jupiter (the Roman version of Zeus) in Gate 37 (family). This position suggests that his Hebe is motivated by wanting to keep his family empire alive. Perhaps it also explains the younger wives, as he seeks to bring his own Hebe (helper) into his family. His wife Jerry Hall has Hebe in a channel with her Earth.

Oprah Winfrey has her Hebe in Gate 63, making a channel with her Pluto. She has a focus on helping (Hebe) to create a new world (63) by experimenting (4) with powerful transformative practices

(Pluto). Her Hebe is a mental pressure that motivates her thinking and planning.

In your Design, Hebe shows your pathway to vitality and youthfulness. Hebe encourages you to pay attention to what truly influences you as a long-term goal. She is a strong indicator of where you desire to be of service in practical and useful ways. If you have a prominent Hebe activation, consider what your "ambrosia" might be.

Hekate

Minor Planet Number 100

Hekate is the crone goddess of the ancient Greeks. She protected against evil spirits and her shrines were found at three-way intersections, a guide for liminal times when we stand at a crossroads, or at a gate beyond which we can see only wilderness. When you are at a threshold moment, not knowing which path to take, Hekate will be your guide.

There are aspects of Hekate hidden in a culture that can only see death as a final ending. Hekate is a necromancer—a shaman of death—drawing out the stories of what is done and bringing renewal to what seems completed. When Persephone was taken, it was Hekate who guided Ceres through the night to find her, holding aloft flaming torches. In the Ovid Hymns she is described as the keeper of the keys to the cosmos. Hekate is your guide when you face difficult and seemingly impossible choices, opening locked doors, shining light, and pointing the way forward.

It's difficult to describe the power and magic of Hekate in a time when the ability to draw something forth from the darkness is considered imaginary or superstitious. Even the guidance of Hekate can seem ill-founded and against rational judgement. Hekate is a soul's knowing of what's complete and what might be coming next.

Hekate's Discovery

The asteroid Hekate was discovered in Gate 13 Line 3. This is a gate of gathering at the point of transition, waiting for the right companions to join you. Together you have the courage to venture into the unknown. In Line 3, Hekate supports us to pause and quietly seek guidance about the correct path. If we push ahead prematurely, we may find we have made the wrong choice, gathered with the wrong companions. These things should never be rushed. We need to feel the subtle influences announcing major change and see where they are guiding us.

Hekate's Mythology

While the asteroid name is Hekate, the more common spelling is Hecate, and occasionally Hecat. She may be the quintessential crone energy, but she contains within her the three ways—the triple aspects of the moon goddess, the maiden, and the mother and crone. Originally a protective and wise energy, patriarchal influences changed her character to one of manipulation and black magic. There was much to be feared when the crone announced her presence! No longer bound by the rules of daughter and wife, no longer owned by father or husband, Hekate was free to wander the mountain passes, guiding the lost to their next true path.

Hekate in your Human Design

Jane Fonda has Hekate in Gate 28, with her Vesta, letting us know she is dedicated to learning from her experiences by following Hekate as she takes on the big picture issues (Gate 28). She has Pallas in the same channel in Gate 38, which gives her a strategic brilliance in bringing people along with her, helping them see her unique perspective.

Hekate shows your process for making the crossing from one stage of life to the next. When transiting planets pass over your Hekate you know you have choices to make. Through Hekate we gather wisdom through experience and revelations about where we go next.

Hephaistos

Minor Planet Number 2212

I've always been intrigued by Hephaistos. Born lame, rejected by his parents, thrown off Mt Olympus, brought up by strangers on the island of Lemnos, refused by Pallas, cheated on by his wife Aphrodite, life was tough for Hephaistos! The one redeeming feature of his life was his capacity to create beauty and magic.

Hephaistos' Discovery

Hephaistos was in Gate 25 when he was discovered in 1978. Heracles, the hero, was also there, which is interesting as I see Hephaistos as a kind of anti-hero.

Aphrodite turns up in Gate 53, which seems unconnected to Hephaistos, which makes sense as despite their marriage there seems little love between them.

His lameness was likely a metaphor for Hephaistos' true affliction, his inability to truly accept the value of his own gifts. He always wanted to be Hercules, never the master craftsman. He longed for the arduous trials and adulation of the hero, not the quiet satisfaction of the forge.

His position in Gate 25 Line 5 describes how he dwelt on his affliction in such a way that he became more and more entangled in an identity that didn't seem to fit. If he had seen himself as the hero-craftsman he truly was, if he had allowed true love to find him rather than projecting his need for affirmation onto his relationships, Hephaistos would have been transformed.

Hephaistos' Mythology

Hephaistos was the Olympic god of fire who crafted weapons imbued with special powers. He was also into robots and built automatons to help him with his work.

Hephaistos was known for his ingenuity and inventiveness. He created many of the iconic items of Greek mythology including Hermes' winged helmet and sandals, Aphrodite's girdle, Athena's breastplate, Helios' chariot, all the thrones on Mount Olympus, and Eros' bow and arrows.

I found Hephaistos in a significant position in the charts of Elon Musk, Steve Jobs, and Nikola Tesla. It's clear this asteroid has some influence in regard to technology and artificial intelligence, particularly where the level of innovation seems god-like, where it pushes the envelope of our idea of what life is—iPhones, self-driving cars, free energy.

I often find a significant Hephaistos activation in the charts of people who create or work with what we might call tools of consciousness, for example Human Design.

There's another, more personal, aspect of Hephaistos and we can see that in his relationship with Aphrodite. Despite being a master (one could say divine) blacksmith, Hephaistos was not lucky in love. He was besotted with Pallas Athene, who wanted none of it. He married Aphrodite, who much preferred Ares. After coming upon his wife with her lover Ares, Hephaistos demanded her bride price be returned. He doesn't see his own value and so his life becomes transactional.

Hephaistos will never be the handsome hero. Limited by his lameness and with a history of rejection, he has to learn to work with what he's got. He projected his love of beauty onto Aphrodite and his skill at crafts onto Pallas. His rejection by both women reflects his inability to own those aspects of himself, perhaps seeing them as not manly enough.

Hephaistos is the same archetype as the asteroid Vulcan and the Uranian object Vulkanus.

Hephaistos in your Human Design

I looked at Robin Williams's chart because he was very successful but wasn't your normal handsome movie star. Robin Williams had Hephaistos in Gate 20, with his Neptune in Gate 57, and Earth and Chiron in Gate 10. This channel is about an individual learning to love and appreciate themselves. His movies are often about learning to understand, love, and embrace those who are a bit different.

In your Design, Hephaistos can represent where you have a high level of skill for creating tools that are beautiful, futuristic, and practical. There is a psychological component to Hephaistos that requires we let go of some idealised version of who we think we should be. When we stop seeing this as a personal failing or affliction, we can embrace our new sense of self and the nature of our true gifts.

Hygeia

Minor Planet Number 10

Hygeia gives information about how you can best maintain a wholistic level of health. A significant Hygeia indicates a focus on health, often as a career.

Hygeia is the fourth-largest asteroid and meets many of the requirements to be reclassified as a dwarf planet candidate. My experience is that she is quite a straightforward archetype without the layers of complexity we find in many of the asteroids.

Hygeia's Discovery

Hygeia was in Gate 18 when she was discovered in 1849. This gate deals with remedying situations when things have degenerated or become spoiled. Also, in that gate we find Veritas, the goddess of truth. So, we know Hygeia's remedy must include a return to truth. In the adjoining Gate of Joy (58), we find Pallas, who was also known as a goddess of healing and whose statue stands alongside that of Hygeia on the Acropolis in Athens.

Hygeia's Mythology

Hygeia comes from health royalty. Her grandfather was Apollo, her father Asclepius and she is mentioned in the Hippocratic Oath.

"I swear by Apollo the physician, by Aesculapius, Hygeia, and Panacea, and I take to witness all the gods, all the goddesses..."

Hygeia represents the ways in which we maintain holistic wellbeing. Her sisters Epione and Iaso were about recuperation, but Hygeia is about prevention, particularly when it comes to hygiene in all its forms. She is about standards of sanitation, for example handwashing before surgery.

As the only female in the ancient health triad, with Apollo and Asclepius, she represents women's health, the caring professions, ensuring caring (health care) for all, and a more intuitive approach to health.

But Hygeia goes deeper, to the relationship between the unseen and the seen. Not just bacteria, but also the subconscious and its effects on health. The word sanitation has the same root as sanity and

Hygeia was prominent in the charts of both Freud and Jung who unveiled the world of the unconscious and its effect on health.

Hygeia in your Human Design

Louise Hay had Hygeia in Gate 64 with Vesta (what she is devoted to) in the adjoining Gate 47 with Elatus (writing skills). Louis Pasteur, known as the founder of preventative medicine and who discovered pasteurisation which destroys harmful bacteria in food, had Hygeia in Gate 41 with his North Node (Destiny).

You may find that if you have a prominent Hygeia in your Design, you have a particular interest in health as a profession, and that your interest tends towards the effect of what are usually unseen influences.

Your Hygeia activation shows what you can best do to maintain your health on all levels—mind, body, and spirit. Think dream journals, somatic healing, and archetypal work, with a focus on prevention.

Interamnia

Minor Planet Number 704

Interamnia is the Greek word meaning "between rivers." Although you've probably never heard of it, Interamnia is the fifth largest asteroid in The Main Asteroid Belt, which gives it some importance.

It represents a creative space where something will eventually come into form on its own terms and in its own time.

Interamnia's Discovery

Discovered in 1910 when it was in Gate 3, Interamnia represents new beginnings and potential coming into form. She represents the creation of something that will last, there is no need to rush things or impose your expectations on it.

This suggests allowing the correction of old patterns (Sun and Mars in Gate 18) to bring a new partnership of the highest resonance (Neptune in Gate 62) of the universal mind (Uranus in Gate 61). Emotional clarity (Mercury and Venus in Gate 6) is important. New ideas arise suddenly when we allow the time and space for our own explorative journey.

There is no mythology for Interamnia.

Interamnia in your Human Design

John Lennon had Interamnia in Gate 57, with his Personality Sun and Design North Node, this is a gate of participation. Lennon was deeply engaged with the music he created. Albert Einstein had his two Interamnia activations on his Uranus and Neptune. He was known for allowing paradigm-shattering ideas to form in his mind.

Your Interamnia is a place where you need to step aside from trying to force outcomes. Your role here is to recognise something coming into being and hold a space for it. There is flow happening (two rivers) and it needs time for you to become the person who can receive and work with the new creative ideas.

Isis

Minor Planet Number 42

Isis is a bridge between the all-encompassing Great Mother and a more democratic approach to spiritual experience. Through her, we can touch the power of the gods and take creation into our own hands. Isis has powers of restoration and finding the missing piece of the puzzle that brings something magically to life.

There is also what's known as a *hypothetical* called Isis-Transpluto which has a completely different meaning than the asteroid Isis.

Isis' Discovery

Isis was in Gate 34 when she was discovered in 1856, with the Sun (in Gate 20) and Earth (in Gate 34). Both Hephaistos and Vulcan are in Gate 20, suggesting superior skills to create new tools of consciousness. It is quite an extraordinary discovery chart, with three channels defined, including Mercury (the messenger of the gods) and Neptune (higher possibilities) activating the Emotional Manifestor Channel 22/12!

In Line 5 in the Gate of Great Power, Isis represents releasing your grip on how you think things should be, working with changes in fortune and unleashing the new opportunities they offer.

Isis' Mythology

Ancient Egypt was strongly matriarchal until around 3,000 BCE when the pharaohs were established as kings. Isis was enthuseiastically adopted by the ancient Greeks and Romans, her cult being one of the most popular mystery schools in Roman society. There is little remaining evidence about the rituals of Isis. Initiates were sworn to secrecy.

In Egypt, these rites were only undertaken by high-ranking priests but once they reached Greece and Rome, they became more widely available. There is an empowerment in this, as more people were able to make contact with the experience of their own inner world and meet the gods personally.

There are many stories in the myths related to Isis. One of the most important is of her search for the body of her murdered husband Osiris. Isis discovers the hacked-up pieces of her husband's body on the banks of the Nile and gathers them in a basket. As she reassembles Osiris, her grief and sexual desire stir him to life, allowing her to have sex with him and conceive their son Horus.

In one version of the myth, Osiris' penis is said to have been eaten by a fish, so Isis creates a golden phallus to enable her to be impregnated with her son.

The mythology of Isis comes out of the old story of the goddess as kingmaker. In hieroglyphics, her name is the same as THRONE. She was the basis upon which the king was able to rule.

Isis in your Human Design

Isis had magical powers to restore the dead by stimulating a desire for life, particularly through sexual desire, stirring us to action. It makes me think of someone who has been alone too long and suddenly meets someone and comes to life. There is a deeper sense of creation in Isis, coming into contact with the universal creative principle and using it to create something that was previously missing from our consciousness.

Einstein had Isis in Gate 11.6 and his Design Sun in 11.5. There's no doubt he found a missing piece of the puzzle! This is the Gate of Heaven meeting Earth and home of the Galactic Center. Einstein's theory of relativity changed our relationship to reality.

Greta Thunberg has Isis in Gate 45, with her Design Earth in Gate 21. Gate 45 is about leadership, what we invest in, and how we create communities. In Line 5, Thunberg's Isis is creating a new kind of leadership that draws people together on a global level to act out shared values.

In your Design, Isis represents the way you can stir something to life. This often requires surrendering to changing fortunes, allowing the missing piece of life's puzzle to reveal itself.

Lilith

Minor Planet Number 1181

There are four variations on Lilith, but only one of those is an asteroid, the others are theoretical points. Each face of Lilith has its own characteristics, although the underlying themes are similar. We'll be talking only about the *asteroid* Lilith, which represents the rage of misrepresentation and a refusal to be part of something that seeks to include you only on the condition that you give up something essential of yourself.

Lilith's Discovery

Discovered in 1927 when she was in the Gate of Influence (Gate 31 Line 6), it's fitting that Lilith continues to hold us in thrall. This is a gate that wants us to recognise influence as a mutual relationship. I can't manipulate you to do anything that you didn't want to do in the first place! I find this so interesting, given the blame that is heaped on Lilith for things like forcing men to have orgasms in their sleep. Line 6 is about being committed to those things that influence you, not just being all empty talk that rejects authentic exchange. How might we have been influenced if we had entered into a real conversation with the demons of our night? What dark desires might we have found if we stopped blaming the influence of others for the actions we wish we hadn't taken?

Lilith's Mythology

Lilith was an ancient demon spirit, winging her way through the desert nights, unruly and untamed. She was considered particularly dangerous to women and children. Given the opportunity to be Adam's first wife, Lilith impolitely declined. Compared to the goodness of Eve, Lilith came to represent all that was wicked and eternally damned.

Like so many of these dark feminine archetypes, Lilith is the representation of the irrational, the instinctive, the primal, and the nocturnal. I approach Lilith as being the place in our Design that says, *I will not submit to any system which refuses to acknowledge me as a full participant.*

Lilith in your Human Design

Hillary Clinton has Lilith in the same gate as her Moon, which explains much of the demonising she experiences. I am struck by the synchronicity of her being accused of kidnapping and killing children. This is such a fundamental part of Lilith's mythology, and it shows that the Lilith archetype has some role to play in our society's inability to address a whole range of harms to women and children.

Grace Tame, voted the Australian of the Year in 2021 for her activism on behalf of child sexual abuse survivors, was pilloried recently in the Australian press for not smiling at the Prime Minister while attending an official function. She commented afterward that abuse survivors are expected to be submissive, to smile for the benefit of the men who abuse them. Tame is often fierce and angry and is much loved for speaking so directly on behalf of survivors. She calls out the unquestioned assumption that survivors should stay silent in case their stories impact negatively on what many see as otherwise blameless abusers.

Tame's Lilith is in Gate 5 (Waiting) with her Personality Mercury (communication). She is expressing (Mercury) her Lilith through a process of waiting patiently for the right circumstances. When the opportunity presents itself, Tame's integrity shines like a beacon, calling out the marginalisation of women and in particular child sexual assault victims.

In your Design, Lilith shows where you will not surrender yourself to any system, organisation, or person who seeks to objectify you and blame you for their own misdeeds. Lilith can trigger rage and sorrow at the depth of loss experienced both on a personal level and collectively. In her constructive phase, Lilith will speak and act with a fierce integrity.

Kassandra

Minor Planet Number 114

Kassandra represents a natural intuitive gift to see what might go wrong, a kind of early warning system. It can be a place where you feel people don't take you seriously or fail to heed your advice.

Kassandra's Discovery

In her discovery chart, we find Kassandra in Gate 30, Clarity. another seer, Okyrhoe, is in Gate 41, making a channel with Kassandra. It makes me think about Apollo's role in usurping the Delphic Oracle, talking the voice of the goddess who had traditionally spoken truth from that place. In the chart Apollo is in Gate 6, conflict.

In Gate 30, perhaps Kassandra's oracular gifts are alerting us to preserve the things to support our brilliance. In Line 1 we are waking up to our clarity, to finding a way to walk our brilliance into the world. There is one defined channel in the discovery chart, made up of Gates 57 and 10. This is about deep participation in life, walking close to the powers of our own creative brilliance. Kassandra reminds us to trust those powers and to listen closely as they guide us.

Kassandra's Mythology

While the usual spelling is Cassandra, the asteroid is named Kassandra.

Kassandra was a beautiful woman with a powerful gift of prophecy, said to have been bestowed on her by Apollo in return for agreeing to become his consort. But Kassandra, for reasons we don't know, decided to refuse Apollo's advances. Perhaps she didn't agree in the first place and the arrangement was all in Apollo's imagination. Perhaps Kassandra changed her mind. Apollo was used to getting everything he wanted, but couldn't take her gift away, so he laid a curse on Kassandra. From that moment on, Kassandra's predictions were met with disbelief.

The conflict between Kassandra and Apollo can be seen playing out in relationships and in the larger society, where feminine intuitive knowing is considered hysterical, irrational, and unreliable. Apollo was a god of light, reason, and rationality. We can read Kassandra's story as a rejection of the "irrational" intuitive voice but look carefully at what triggered Apollo's curse—Kassandra's rejection of his sexual advances, her breach of the (apparent) bargain that we are not even sure she had any part in making.

Kassandra in your Human Design

Author and scientist Rachel Carson had Kassandra in a channel with her Pluto. Carson wrote *Silent Spring* in 1962, alerting people to a new concept, that pesticides didn't just stay on weeds and pests, but spread through the entire ecosystem. She was widely ridiculed and criticised, being called a communist and a fanatic defender of the cult of the balance of nature.[15]

In your Design Kassandra shows where you have a natural intuitive gift. It can show where you have experienced disbelief, where your own intuitive truth struggles to find a place in your reality, or in the way others see you.

You can explore the relationship between your own Kassandra and your Apollo activation. How does the position of your Apollo represent where you favour a more rational approach to life? How might that part of you override your more intuitive aspects, as represented by your Kassandra activation? How could these two aspects—the intuitive and the logical—work together to create a more fully realised view of your life?

---

[15] https://en.wikipedia.org/wiki/Rachel_Carson

Lucifer

Minor Planet Number 1930

Lucifer conjures up ideas of an evil trickster, tempting us to the dark side. But there is so much more to this archetype than his popular devilish reputation. We can understand Lucifer better if we reclaim his darkness from those who proclaim it evil and recover our capacity to go into the dark to find new light. His warning to us is not to get trapped in the darkness of being victimised and of listening to our ego over what we might call our better self.

Lucifer's Discovery

Like many others, the asteroid Lucifer was discovered in Gate 3, which teaches us about the difficulties of creating a new kind of order from the chaos of decaying old patterns. Aphrodite (the Greek version of Venus) is in a channel with Varda (goddess of the stars) and Lempo (Finnish goddess of fire). Saturn is in the Gate of Brilliant Fire (30) with Astraea, another goddess of the stars. So we have a lot of evidence pointing toward being guided by clarity, light, and the stars.

Venus however, in the Gate of Conflict (6), suggests a tendency to see relationships as a zero-sum game. You win, I lose. Her position in Line 6 guides us to an understanding that this kind of thinking suppresses personal growth and limits our capacity for intimacy.

Lucifer is joined in Gate 3 by Admetos, representing breaking through blockages to create deep transformation. Lucifer's Line 2 placement says we shouldn't take the easy way out. We have to wait through the night to discover what new possibilities the dawn holds for us.

Lucifer's Mythology

The name *Lucifer* can conjure up ideas of Satan and fallen angels, but he has a more positive face and has long been associated with the planet Venus. The name Lucifer comes from the Hebrew, meaning light bringing; Venus often rises before the Sun, signalling the dawn. There are also some intriguing links with the Book of Enoch, an ancient Hebrew apocalyptic religious text, which explains how Lucifer used his free will to disempower rather than to light the way.

We can find two aspects of Lucifer. Before Christianity, the Romans associated Lucifer with Aurora, goddess of the dawn. Although he wasn't a god, he was said to order the heavens. This is the light motif.

The other motif is the dark angel, linking to the ancient rites of descent to the underworld found in the ancient mysteries of Innana and Ishtar. These rituals were a reconnection with the power of shedding the ego and meeting the dark self, represented by the earth, the mother, the womb, and the grave.

Lucifer in Your Human Design

I was trying to think of a Lucifer type and wondered if he would show up in the Design of nice actors who play dark characters. Idris Alba, the villain in *No Good Deed, Beasts of No Nation*, and *Star Trek Beyond*, has Personality Lucifer in the same gate as his Uranus (57), and his Design Lucifer (in 47) makes a channel with his Personality Sun (in 64). Alice Cooper, a singer with a reputation for being a beautiful kind human being with a dark stage persona, has Lucifer in Gate 14, making a channel with his Earth in Gate 2.

Author Jane Austen has Lucifer in Gates 44 and 26. She has her Venus in Gate 44. Think of her character Mr. Darcy in *Pride and Prejudice.* Was he prideful? Or was Elizabeth Bennett too prideful to see his true nature? The journey through a certain kind of isolating social darkness requiring a reckoning with one's own nature, is a common theme in Austen's novels.

In your Design, Lucifer represents a place where you may try to avoid facing your prideful self and end up taking a tumble into the dark side of your own nature. Once there it's best to accept the slow and deep pathway back to a new dawn. You might discover things you don't like about yourself, but you will be rewarded with a new level of personal integrity and a greater trust in yourself.

Magdalena

Minor Planet Number 318

I was delighted to find an asteroid called Magdalena. In my work with her over the years, I have found she represents an awe of the divine gifts that women in particular bring to society.

Magdalena's Discovery

Magdalena was in Gate 49, Revolution, when she was discovered in 1891. She was joined there by Ceres and Eros, suggesting a powerful conjoining of erotic desire with the notion of what nourishes us. Venus is in the same gate as the Sun, so we have that idea of eroticism and desire as part of our personal development. Aphrodite (a life of desire and pleasure) is in Gate 50 (a new era) with Uranus (revolution).

Magdalena is a touchpoint for the inclusion of deep feminine wisdom and how that will support a profound shift in culture.

Magdalena's Mythology

According to church orthodoxy, both Mother Mary and Jesus were virgins. They were both, to coin a phrase, immaculate. But not Mary Magdalene, who was for centuries described as a prostitute. This says less about the nature of Mary Magdalene and more about the terror felt by early churchmen when confronted by their own sexual desire.

The early religious and cultural superstructure of Ancient Greece and Rome denied any very active role for women. The trinity, entirely masculine, forcefully replaced the triple goddess of mother, maiden, and crone. There is a certain insanity in denying women's creative role, but Magdalena takes us even further into this falsity.

The early church created a spiritual tradition devoid of a feminine voice. The intimacy of women's sacred experiences has been marginalised, and its connection with the dark, the body, the womb, with pleasure and sexual desire were all suppressed.

Magdalena in your Human Design

Princess Diana has Magdalena in Gate 24, with her Ceres. This is a gate of returning again and again to your core self in order to find your true direction.

Serena Williams has Magdalena in Gate 57 with her Jupiter, and Zenobia (warrior queen) in Gate 34. She also has Lilith in Gate 10. There's a lot of female power there, and Williams has been instrumental in getting the women's tennis game on par with the men's. Magdalena can influence us to seek inclusion and equality in more mundane ways as well.

In your Design, Magdalena shows your pathway for recovery of the feminine aspect of the sacred. A prominent Magdalena indicates you have a role as a spiritual teacher, focusing on particularly feminine qualities. The asteroid Magdalena carries with it a devotion to the felt sense of divinity, the ineffable sense of belonging to a greater story, and a knowing of the connection of all things.

While there are many archetypes that represent the feminine aspect of divinity, Isis for example, Magdalena seems to carry a sense of mission to restore our relationship with what author Margaret Starbird calls the left hand of the sacred.

Medea

Minor Planet Number 212

Medea represents a great gift that may be squandered on the wrong people. When she finds her crew, Medea can find a solution to even the most difficult of problems.

Medea's Discovery

When she was discovered, Medea was in Gate 59 Line 5 with Salacia (being seduced into giving up everything for romance and love) and Aphrodite. This line is described in the Rave I Ching as the femme fatale! It's also about learning how to take the reins of leadership when your people are in unknown territory. In the adjoining Gate 6 we find Europa, who also carries a theme of being seduced away from her own life, although in her case it's not the opportunity to contribute to a greater cause, but to find security. This is a channel of intimate sexual relationships where we can rush into bonds too quickly, offer up our expertise and find our own life washed away. It's not just in relationships and can apply to any creative venture, job, or business partnership.

Medea's Mythology

Jason and the Argonauts were hailed as heroes after undertaking a mission to recover the mythical Golden Fleece. But it was Sorceress Medea who helped Jason navigate through every challenge on his impossible quest.

Medea is often portrayed as a power-hungry evil sorceress who murdered her brother and her own children. This is the kind of makeover given to many a powerful feminine archetype by the ancient Greeks. Medea was in love with Jason, but even more than that, she was excited to be part of something, sailing out with the Argonauts to recover the Golden Fleece.

They would never have retrieved the Golden Fleece but for Medea and yet she was vilified for seeking recognition for her part in the Argonaut's success. Think of Hermione in the Harry Potter series. Harry is always the hero, Hermione the clever girl who was too big for her boots.

Medea sits at that crossroads, where women are instrumental and yet still secondary where their work is claimed by another. She is represented as having great intelligence and skill, traits typically attributed to men in Ancient Greece.

Medea in your Human Design

Princess Diana had Medea in Gate 45, making a channel with her Design Sun. She often allowed her emotions to override her reason. Her passion and longing for love were her undoing. But we also see how she was beginning to apply her gifts to more universal causes, like land mines and AIDS which perhaps aligned more with her own life purpose.

Beyoncé has her Medea in the Gate of Limitations (60) with her South Node, suggesting she can always find solutions to limiting situations. The South Node is something we are naturally skilled in and can easily rely on. The asteroid Europa in adjoining Gate 3 brings the energy of an Empress or Queen into the mix. Her Medea is focused on the power of her presence as a kind of mix of celebrity and nobility. She has Terpsichore (the dancer) in Gate 4 making a channel with her Earth and Orpheus (the musician) in 63. This suggests to me that the passion of Beyoncé's Medea is more about her image and influence than about her creativity which is in unrelated circuitry.

In your Design, Medea represents allowing passion and a desire for love and acceptance to overrule our reason, so that we rush into bonds before we are sure they align with our own greater purpose. When we do this, we often find we are tolerated but not really included, and that others take credit for our work.

Medea carries a gift of finding solutions when others are all at sea. Gate 59 tells of those disorienting times when the familiar is being washed away. Medea can always find a way to move through the uncertainty and crisis to successful outcomes.

Medusa

Minor Planet Number 149

Medusa is the death goddess, the monster who petrifies anyone who glances her way. Beheaded by the hero Perseus, Medusa haunts our imagination with her many-eyed serpents and horrifying gaze. But who is Medusa really? And why are we so fascinated with her?

Medusa says a lot about how we relate to, manage, and recover from trauma. In your chart she also represents powerful aspects that you may have difficulty expressing, even perhaps projecting them onto others.

Medusa's Discovery

Mostly, when I look at the discovery charts for asteroids the meaning jumps out at me. Not so for Medusa's chart. She was in Gate 22 Line 3 when she was discovered in 1875, and there is nothing much else in that channel or even surrounding her in any meaningful way. It made me think about how isolated the mythological Medusa was. Perhaps it was her lack of connection that led to her transformation.

However, there is a real synchronicity in her position in the chart, no matter that no one else turned up to meet her there. Gate 22 is all about our relationship to beauty. Line 3 teaches us not to get seduced by surface beauty, but to understand both the inner and outer person.

Perhaps a significant Medusa in our Design may have us clinging to external beauty out of fear that without it, all we have to offer is an ugly monster?

There are two defined channels in her discovery chart, suggesting she is part of creating a more caring world where people are better able to connect authentically.

Medusa's Mythology

The first we hear of Medusa is as a beautiful and graceful young maiden who was raped by Poseidon in the temple of Athena. Furious at the desecration, Athena punished not Poseidon but the "temptress" Medusa. She turned Medusa into a monster and confined her to a cave at the edge of night. Anyone who looked at the monstrous Medusa was turned to stone.

Suddenly, through no fault of her own, Medusa went from a happy young maiden to a dreaded monster.

Perseus is one of the great heroes of Greek mythology because he vanquished the monster Medusa. He was only able to do so with the help of Athena and by using Medusa's power against her. Yet even after her beheading, Medusa's power was so great that Perseus had to keep her head in a sack, only pulling it out occasionally to stun his enemies. Perseus repaid Athena for her help by giving her Medusa's head for her aegis, her shield; She took Medusa's power and used it for her own purposes.

We can see a split in the feminine here. Athena gave up all her own femininity to be a mascot for the patriarchy. Separated from her mother and uprooted from her protective group, Athena had no choice but to play along with the violent misogynistic narrative. She collected powerful feminine archetypes like Medusa and transformed them into something more conducive to the predominant reign of Zeus.

Medusa is the beautiful young temptress, her warrior nature destroyed not just by sexual violence, but also by being punished as if the rape was her fault. The story of Medusa is about normalising the blame of women for violent acts perpetrated against them, making use of the wisdom and power of women while claiming to be the hero.

So who was this powerful young woman, and where did this power come from? Medusa was a Gorgon, probably from the Libyan Amazon tribe of female warriors and the serpent goddess Anatha. The serpent goddess motif represents the feminine death energy at play here, the deep transformation requiring us to release the old life. Medusa is one of many feminine archetypes who represent the loss to us of the natural cycling energy from our culture, and the resultant terror of change.

Men evoke Medusa when they want to demonise women who step beyond their allotted place. She represents the parts of the feminine that cannot be integrated into the patriarchal order—Medusa the goddess of death, Medusa the beautiful temptress, and Medusa the warrior. There is definitely a whiff of gaslighting coming at the Medusa woman; she is the perpetrator rather than the victim, she is incapable of wielding power on her own behalf and must surrender it to the hero, she is stony-faced rather than authentically trauma-tised. Medusa is complex and fascinating. Her wisdom and power represent one of the great threats to patriarchal culture.

The paradox of Medusa is that she has been slain by the hero and yet retains such power. The enduring fear of Medusa reveals the paradox at the heart of patriarchy, the fear of death, the fear of women, and the need to demonise women, lest the power of sexual desire and the rage at its denial engulf and destroy everything.

Medusa in your Human Design

Oprah Winfrey has been an outspoken advocate for survivors of sexual abuse and has provided a platform for their stories. She has Medusa in Gate 46, with her Design Mars, a very positive and active energy to reclaim our bodies from the effects of trauma. The use of the word *survivor* is very Medusa, focusing as it does on the power to overcome and even flourish, despite difficulties.

In your Design, Medusa can represent periods of intense change, when we are challenged with new circumstances. We can't go back, so how will we go forward in these new circumstances? Who may have used your power against you by casting you as the perpetrator rather than the victim?

If you have Medusa in a significant place in your Design, you are naturally skilled at helping people face these challenges.

Nemesis

Minor Planet Number 128

When I mention Nemesis in a reading people tend to shudder slightly. There's no need to be afraid of her. The winged goddess of retribution is all about karma. She is just as likely to distribute good as evil. In your Design she is the archetype of alignment with universal law, showing what old karma might trip you up again and again.

Nemesis' Discovery

The asteroid was discovered in 1862, in Gate 16 Line 2, which carries the theme of using your enthusiasm in alignment with your true purpose, not trying to hold onto what's not yours, including power. You should allow what is not yours to flow out of your life rather than trying to hold onto it.

The Sun and South Node are in Gate 34, which is often the energy of pushing against the flow of life rather than having the humility to use your power in service to a larger creative story.

When exploring your Nemesis, check the position of the asteroid Hybris as well, it shows where you are likely to have a bit too much belief in yourself.

Nemesis' Mythology

Nemesis is one of the Moirai or goddesses of Fate. Operating above even the gods, the Moirai were said to appear seven days after a child's birth to accept the infant and to prophesy its fate. This may go back to earlier times when wise women checked the health of an infant and assigned its role in the community based on divination.

The role of Nemesis was to ensure everyone, particularly the gods, lived in a way that was harmonious, balanced, and within the universal principles of good order. When we become hubristic, we are failing the greater purposes set for us by the Fates. Nemesis was said to punish hubris, delivering a dose of humility to those with too much belief in their own good fortune.

Nemesis is common in storytelling, where we assume that in the end the hero will be rewarded, and the villain will be punished. The

Nemesis character is the ultimate archenemy who can't be vanquished. Think of Voldemort in the Harry Potter series.

Nemesis in your Human Design

Nemesis was active in the 2016 US presidential election. Hillary Clinton's Nemesis is in the same gate as Vladimir Putin's Hybris (hubris), and Trump's Nemesis is in the same gate as Clinton's Design Mercury. There was some big karma going on there!

In your Human Design, your Nemesis can turn up when you believe your own personal goals are more important that universal flow, when you perhaps have a bit too much hubris and belief in your own invincibility. Nemesis will turn up in the form of some obstruction in your life, or a person who seems to stand in your way. Her message is to look at how you have created this blockage and she brings a message for how you can move back into harmony with the greater flow of life.

Ophelia

Minor Planet Number 171

Ophelia is the most popular of Shakespeare's heroines, the daughter of Polonius in the play *Hamlet*. Originally viewed as a young woman driven mad by her sexual longing, modern feminists saw a different Ophelia, a young woman seeking to be heard.

Ophelia is the daughter who holds the potential for new life. But she gets ignored by the apparently powerful men who are busy politicking, desperately trying to find a short-term solution. In the end, Ophelia dies, carrying the child who represents a new start.

But it doesn't have to be this way!

Ophelia's Discovery

Ophelia was in Gate 7 when she was discovered in 1877, suggesting she is trying to make a life according to someone else's rules. This is the Gate of The Army, where there is a tendency to march to the beat of someone else's drum and find you have arrived at a destination that ultimately disappoints. In Line 1 we are at the beginning of that journey, and Ophelia asked—are you sure these are *your* goals? Are they meaningful for *you*? Or have you become caught up in the general hubbub or busyness?

Eurydice turns up in Gate 31, Influence, making a channel with Ophelia. The most common myth of Eurydice is found in Orpheus' attempted rescue of her from the underworld. However, mythologist Barbara Walker explains that Eurydice was originally the Underworld Goddess who received the souls of the dead, also related to Queen Heurodis of England. It would have been Eurydice who rescued Orpheus, the myth has been turned on its head! This makes sense. It is Ophelia who has within her the means to solve problems, to change situations. She doesn't need rescuing, only recognition of her solution.

This is accentuated by the position of the Sun in Gate 61—Inner Mystery, offering emergent new big sky solutions to old problems. The Channel of Logic (63/4) is defined, suggesting an ability to see newly emerging future trends.

Ophelia's Mythology

Ophelia doesn't fit in the Greek or Roman pantheon at all, and so there is no mythology as such. But she has over the years developed an iconic status that creates its own mythology.

There is a subtext in Shakespeare's *Hamlet* suggesting Ophelia was pregnant, and that after her father is murdered, Ophelia is alone with her problem. Perhaps it's not an excess of desire and feeling that drives Ophelia to suicide, but the irreconcilable situation she finds herself in. Finding yourself pregnant after being dumped by your beloved, someone you assumed you would spend your life with, is an age-old tale that is as relevant today as it was in renaissance England.

The archetype of Ophelia could represent a woman who becomes a sexual plaything for powerful men, particularly in those circumstances where young women find it impossible to speak out. Ghislaine Maxwell, accused of sex trafficking young women, has Ophelia in a channel with her Earth and Mars. It has been reported that she was in thrall to her powerful father Robert Maxwell, and this may have given her a predisposition to take on the same kind of consigliere role with her business partner, the late Jeffrey Epstein.

I usually work with Ophelia in a broader context, where she represents the process of disowning what we have been unable to explain to others. What do we do when we are approaching a kind of creative birth, gestating within our own feeling nature, and no one around us gives it any credence? How do we acknowledge our own unique solutions and goals when others are marching to a different beat entirely?

Ophelia in your Human Design

In your Design, Ophelia represents personal problems that go unseen, originally by others and perhaps then by ourselves. Those problems hide a more subtle creative process that is difficult to communicate, and which we may have learned to drown out with the apparently more important busyness of the world. Your Ophelia shows where you bring solutions that others may not immediately be able to see, but which you should never give up on.

If you tend to push yourself to busyness, thinking this is how to make progress, go and spend some time with your Ophelia.

Pandora

Minor Planet Number 55

Pandora is generally known for releasing evil upon the world through her curiosity. Taking the lid off the jar, she realised her mistake too late! In astrology, the asteroid Pandora is usually said to show where we create chaos through being too curious. But the original ancient goddess was Pan-Dora—all gifted!

Pandora's Discovery

The asteroid Pandora was discovered in 1858 when she was in Gate 21 Line 4, a gate that's about biting through the surface appearance of things to find a deeper truth. Perhaps Pandora's curiosity is a gift of knowing there is something more than what others want us to believe? Line 4 in this gate tells us that if we persevere, we will find a new kind of truth, perhaps not at all what our curiosity expected, but something much better.

In Pandora's discovery chart the Sun is in the Gate of Oppression, the Earth in the Gate of Grace which asks us to find beauty beyond the surface of things. Jupiter (Zeus) is in the Gate of Gathering, which represents the Ruler. Pandora in Gate 21 represents the Manager, the person who manages and controls everything for the Ruler. What happens when Pandora opens her jar?

Pandora's Mythology

The Greek Pandora was created on the instructions of Zeus purely to inflict misery on humanity in punishment for their theft of fire from the gods. Zeus commands the creation of a beautiful yet cunning and evil temptress who will make men's lives a torment, represented by the countless plagues—strife, death, pain, sickness—that she releases from her jar.

Pandora is a kind of Eve, drawing men to sin. Both Pandora and Eve share the quality of curiosity which brings evil to humanity. For Eve it was the apple and the tree of knowledge. For Pandora, the jar and all its plagues.

Pandora is also a motif for the father as the sole creator. Like Pallas, Pandora had no mother. She represents the abstract notion of a pure and perfect creator god, and the imperfection of humanity. Women are the more imperfect gender in this narrative. Despite supposedly being mentally incapable, their curiosity must have been a severe trial for a god who didn't welcome questions. How else to explain away the temptation of men to go back to sin again and again and to then blame the evil nature of women?

Despite being a minor figure in Greek mythology, Pandora is very much alive in our modern imagination. She is an enigmatic figure who can take on the meanings attributed to her from many different aspects of her original story.

Since the time of Pandora and Eve, women have been blamed for not keeping in their place. They have been shamed for being too beautiful, too tempting, too clever, too curious. If women would just stop being so damn attractive! If only they would stop challenging the perfection of god's creation!

If we peek beneath this curse energy, we find the original nature of giftedness in Pandora. She is created from the Earth and represents nature's bounty. The chaos of living in deep relationship with the natural world is seen by the godfather as a loss of control. Pandora holds the cure to the lie that abundance comes from obedience to the one god and instead lies in discovering and developing our own gifts.

Pandora in your Human Design

Greta Thunberg has her Pandora in Gate 57, with her Medea (solutions, gifted sorceress) in Gate 10 (Personality) and Gate 34 (Design). I've always been impressed with Thunberg's ability to cut through the behaviour (Gate 10) and power dynamics (Gate 34) to point out truths that others might miss (Gate 57).

Pandora represents where you have natural gifts but have been led to believe they are a curse and have tried to keep a lid on them. She shows the ways in which we try to keep control of ourselves on behalf of the patriarchy, religious and family beliefs, and where our curiosity naturally leads us to explore a deeper truth.

Persephone

Minor Planet Number: 399

Persephone, the beloved daughter of Ceres, was Queen of the Underworld. One lovely day, when she was out picking flowers in the fields with her mother and friends, Persephone was violently kidnapped and taken to the realm of Hades. In your Design, Persephone brings an understanding of personal initiation– what we lose and what we gain from our dark nights of the soul.

Persephone can show where our own subconscious or soul self can kidnap or trap us, and what we need to do to reclaim our dark power to rule our own lives. She guides us through those transitional moments when one chapter of our life is complete. It is time for something to be destroyed, forcing us to release cherished comfort zones and to begin our journey to master the new place in which we find ourselves.

In particular, Persephone talks about a loss of innocence in the process of gaining wisdom from experience and taking responsibility for making ourselves at home wherever we find ourselves in life.

Persephone's Discovery

The asteroid Persephone was discovered in 1895, when she was in Gate 64 Line 2. This gate is about transitions, the completion of one stage of life leads us to begin wondering—what's next? It particularly speaks to our anxiety about whether we really should leave our safe and comfortable present to launch ourselves into an uncertain new phase of life.

The guidance in Line 2 is that yes, it is a good idea to make changes but don't leap into anything. Big changes can be managed by setting an intention (this is a mental gate, so it starts in the mind) and then choose your steps wisely. It's a sifting time, what to renew, what to release, discovering through your actions the coherence and harmony to align with the new cycle.

I love the synchronicity of the asteroid discovery charts in Human Design. Gate 64 Line 2 talks about a cart rushing to the riverbank at a headlong gallop, with the danger of falling in and being lost forever. The story of Persephone's abduction has her being snatched by Hades as he raced by in his golden chariot, the ground opening up to swallow them both.

When this happens, we shouldn't panic or try to get back to our old life. Instead, we can take some time to explore this new territory and reorient ourselves.

Although as a young maiden kidnapped and taken to the underworld, Persephone must have been disoriented, she took command of her new circumstances, becoming Queen of the Underworld. The Sun is in Gate 55 Line 5, which talks about a beautiful new life, bringing the gift of confidence and independent strength. The Earth in Gate 59 Line 5 lets us know this is a time of transition when we must use all our resources for the crossing, holding nothing back.

Persephone's Mythology

Persephone is the maiden aspect of the triple goddess, the seed of life which must be fertilised in order to grow into its full form. According to Barbara Walker, Persephone was Queen of the Underworld long before Hades became God of the Dead.[16]

Persephone was also known as the Kore, the seed maiden. She was the maiden aspect of the triple goddess from a time when women were keepers of the sacred mysteries of agriculture.[17] The seed maiden openly coupled with the sacred king at the autumn sowing in order to assure a good harvest.

The seed was buried beneath the earth, and this is the basis for the violent kidnap in the more recent story. But there was no kidnap for the seed maiden. This was a voluntary withdrawal in the dark cave-like womb where the seed nestled, waiting for spring. Her name comes from phero and phonos (she who brings destruction). The old life is gone. We are in the dark. We must patiently await the rebirth.

---

[16] ISBN: 0-06-250925-X(pbk)
Barbara G Walker, *The Woman's Encyclopedia of Myths and Secrets,* 1983 HarperCollins, New York.
[17] ISBN: 0-14-017199-1
Robert Graves, *The Greek Myths*, 1960, Penguin Books, London.

This narrative of staying on the surface obscures the power of our inner world and persuades us not to listen to the inner transformer telling us something is done. We need to allow it to be destroyed.

Persephone's mother Ceres rages against her kidnap. She wants her daughter back. She wants things to be as they were. We could possibly say that Ceres doesn't want Persephone to become her own person. I often ask women to look at their Persephone activation as a place where they need to grow beyond their mother's approval, to take full responsibility for their desire not to come back within her sphere.

Long-term estrangement between mothers and daughters can often come from the mother's failure to allow the daughter to be her own person, and the daughter's fear of her mother seeing who she has become. Persephone chose to remain with the bad boy Hades, rather than return to her own life and this choice often needs to be reconciled in our relationships. We can see Persephone in any situation where we have to say to someone, "I chose this new life over my existing relationship with you."

Persephone in your Human Design

Beyoncé has Persephone in Gate 43, along with her Personality Moon, and Chiron in 23. She would experience insights that suddenly change how she thinks about her life, but gradually, as she grows into that Chiron energy, she learns to give life time for the change to play out. This helps her feel more secure in knowing what's true for her, rather than feeling "kidnapped" by her insights.

Persephone in your Design shows the way in which you need to voluntarily become the seed, buried beneath the ground, unsure what you are becoming. In your dormancy, a new power arises within you. Sometimes this asks you to step away from existing relationships, particularly with your mother. You have outgrown something, and relationships will need to be recreated when you come to the surface in your new form.

Psyche

Minor Planet Number 100

Psyche represents your personal pathway to psychological health and wellbeing. Her position in your Design shows the genesis of a long-standing wound to your soul, and the journey you will take to awaken, both to yourself and to love.

Psyche's Discovery

Psyche is the Greek word for soul. Her astrological symbol is a semicircle topped by a star, representing a butterfly's wing, symbol of the soul.

If you want an example of the incredible synchronicity of the discovery charts with the meaning of the asteroids, you can't do any better than Psyche. She was forced to undergo impossible trials by the powerful Aphrodite, and only succeeded through the intervention of supernatural forces.

Psyche was discovered in 1852 when the asteroid was in Gate 29 Line 3. In his book *The Laws of Change: I Ching and the Philosophy of Life*, Jack Balkin describes it like this:

> *You are surrounded by forces you do not fully understand. If you try to act now, you will only make matters worse. You are in more trouble than you think. Do not dig yourself in deeper. Exercise self-control. Stop struggling in vain and try to regain some clarity of mind. You must wait patiently now until a solution to your difficulties becomes apparent.*[18]

Psyche's Mythology

The story of Psyche and Eros was written by Apuleius in the 2nd century Latin novel, *Metamorphoses*. Like most of these archetypes, though, the roots of this story go back at least to 5,000 BCE. This is a story about isolation, vulnerability, and the female hero's journey.

---

[18] ISBN: 978-0-9842537-1-5
Jack Balkin, *The Laws of Change, I Ching and the Philosophy of Life,* P 329-30, 2002, Shoken Books, New York.

Psyche is a mortal but was attracting so much attention for her beauty that Aphrodite began to feel neglected. She told her son Eros to seek revenge on her behalf by making Psyche fall in love with a monster. The plan went awry when Eros accidentally pricked himself with his own arrow and fell in love with Psyche. Not the outcome Aphrodite was looking for!

While this is going on, Psyche's father is trying to find her a husband. Apuleius doesn't explain why the beautiful young woman has no suitors, perhaps she seems too good for them? Too beautiful, too pure, too innocent? Psyche certainly seems all of these things, as her father follows the advice of the Oracle of Delphi and leaves her in a towering rock spire to be taken in marriage by a beast even the gods fear. Here, says Apollo, Psyche shall meet her doom.

When she wakes and explores, Psyche finds a wonderful palace with golden columns and silver ceilings. A voice tells her to make herself comfortable and she is entertained at a sumptuous feast that serves itself. Although fearful and without sexual experience, she allows herself to be guided to a bedroom, where in the darkness a being she cannot see makes love to her. She assumes it is the monster of the prophecy, but still, she promises never to reveal his face.

Psyche's two sisters come to visit and, fearfully persuade Psyche to look at Eros. The sisters had sown such seeds of doubt and dread that one night, Psyche shines a candle on her lover's face. Rather than the feared monster, her candlelight unexpectedly reveals the most beautiful creature she has ever seen.

In her confusion Psyche wounds herself with one of Eros' arrows. Eros wakes and, realising his secret has been discovered, flees. Poor Psyche, so in love, wanders aimlessly, feeling suicidal at having lost Eros. She eventually realises her only hope of redemption lies with Aphrodite.

Finally, Aphrodite gets to take her revenge! She revels in her power over Psyche and sets her four seemingly impossible trials. Against all odds, and with help coming from unexpected places, Psyche is able to complete the first three trials. Her final trial is to visit Persephone, Queen of the Underworld, and to bring back a box of her beauty oil for Aphrodite. Of course, Aphrodite knows that no one ever returns from the underworld, and this is a sure-fire plan to be rid of her young rival.

Psyche makes a plan. She will get to the underworld by leaping off a tower and killing herself. The tower itself stops her from jumping and gives her instructions for how to approach Persephone. It tells her to take two coins for the ferryman and two honey cakes for Cerberus the three-headed dog (a representation of Hekate). Most importantly, the tower says, do not be distracted by those who will seek to lure you away from your purpose!

Once she has the ointment, Psyche makes a fatal error of judgement and decides to take a little for herself, to ensure Eros' love. Psyche opened the box and found it contained the sleep of Hades. Eros flew and rescued her from this sleep of the dead. He approached Jupiter to grant Psyche immortality so they could be married.

Psyche in your Human Design

There are so many layers of meaning to Psyche's stories. We have the unformed young woman, passively accepting whatever her family chooses for her. Her first awakening comes in the palace with what she assumes is a terrible monster, her new husband. But still, she passively does as she is told and doesn't look upon his face. Her two sisters trigger her first real shift in awareness and following their suggestion she goes against her husband's wishes, holding a candle to his face. Expecting horror, she sees Eros! As she gazes on his beautiful face she is pricked by his arrow and falls instantly in love.

We enter the next phase of Psyche's awakening. Eros flees and Psyche is forced to go to Aphrodite for help. This is the first decision she has made alone, and the outcome is uncertain. Aphrodite certainly offered help, but in the form of deadly trials to prove herself worthy of Eros' love.

Ultimately, Psyche is about our own integration of mind, body, and soul. She represents waking up from the passive acceptance of fate, making our own choices about what we desire and trusting we can make our way to love. Help is always available, although perhaps not from the sources or in the ways we are expecting it.

Dolly Parton has her Psyche in Gate 45 with her Design Uranus. If we think about the idea of trials for love, Parton, who's been happily married since 1966, may have had to deal with issues of freedom (Uranus) to be the breadwinner (Gate 45). We could also perhaps think about her having some kind of psychic attunement to rule breaking (Uranus breaks the rules, Gate 45 can be about tribal rules).

Princess Diana also has her Psyche in Gate 45, with her Design Sun in Gate 21. Again, we can see the potential attunement to the rules. In Diana's case, it was more about having to maintain the status quo to find love, but her trials were in seeing her prince for what he truly was. With Psyche in Gate 45, it would have been difficult for Diana to even see herself in any way as separate from the rules that governed the English upper class. As her original passive acceptance shifted, triggered by her desire for real love, Diana had to find her own truth and take control of her own life (Gate 21).

In your Human Design, your Psyche activation will show the shadow side of romantic love, where we expect to stay in a state of bliss without looking reality in the face. It can show how we have a choice to remain with a potential that is unsatisfying or undergo the real and challenging trials that will reveal our own strengths and capacity for true love. She can show the ways in which seemingly traumatic experiences can open us to new ways of perceiving life, particularly when it comes to relationships.

Psyche can also represent psychic attunement, empathy, mental telepathy, or an ability to manifest or work with energy fields.

Urania

Minor Planet Number 30

Urania is the goddess of the skies and the stars. She was the granddaughter of sky-god Uranus and the Muse of Astronomy. She has a gift for symbolism, mathematics, and systems-informed prophesy and can read the movement of celestial bodies to predict the future, cosmology.

Urania's Discovery

When Urania was discovered in 1854, she was in Gate 13, Fellowship. Mercury, who represents access to the higher mind, teaching, and communication in all forms, is in the same channel, in Gate 33. Urania's discovery chart indicates we should never use our gifts of divination and prophecy to confirm our existing beliefs, but always seek to be challenged in our own thinking. This is a channel that reveals secrets and seeks time to understand their true importance.

Urania's Mythology

Urania was known as the Celestial One, and Heavenly. She is closely linked with the qualities of Uranus, and the planet Uranus was at least partly named for her. She embodies the highest qualities of universal love as a function of cosmic meaning.

Urania was one of the nine Muses in Greek mythology; she ruled astronomy and astronomical writing. The Muses were protectors of the relationship between the skills of culture and the cosmos. Urania was responsible for the music of the spheres. Her expertise included science, mathematics, exploring reasons behind events, music theory, conceptualising abstract and symbolic knowledge, and foretelling the future through stars.

Urania in your Human Design

I was curious to see if Urania showed up in the Design of musicians and found her in David Bowie's chart in the same gate as his Personality Moon and Saturn. I looked at Bowie because of his ability to tap into the zeitgeist with his music, to take abstract cultural ideas and make music from them. So it seems Urania may not just be about astronomy and astrology.

On the other hand, I looked at the Design of Mike Brown, the lead astronomer of the team that discovered the dwarf planets back in 2005. He has his Personality Urania in Gate 31 with nothing else happening there. He does have Ariadne in Gate 7, and she represents a kind of genius level of skill, but I would have expected to see a planet in that channel. This indicates to me that Urania is much more than the mathematical aspects of astronomy. She is about how we make meaning from the symbolic world.

In your Design, Urania represents how you are tapped into the music of the spheres and have an ability to make meaning from abstract and symbolic information, particularly when it has a technical aspect like astrology or music. This is a place in our Design where we are more serious than frivolous, where we have a natural desire to understand the symbolic processes in our lives at a cosmic level.

Conclusion

Adding the asteroids into your study of Human Design changes everything. They breathe life and dynamism into the fixed structure of that one single chart.

The great challenge of exploring the asteroids is that there are, quite literally, hundreds of thousands of them. Where to start? When to stop?

This book lays out a way to enter into the journey of discovering the most significant asteroids in your Human Design. You will find these archetypes feel familiar, a part of you. It's likely you will find their stories showing up in your own life—Artemis walking the forest with her hounds, Medea gradually learning to share gifts with those who appreciate them, Aphrodite being seduced by all that is beautiful.

You cannot "learn" the asteroids and then move on. They are always with you, living through you, teaching you, leading you more deeply into a true expression of who you are. They will not be rushed, so take your time. Pay attention when they call you!

Find your own way to communicate with these mythological energies. You might feel more comfortable with a visual response, maybe drawing or painting what resonates for you. Journaling is a popular way to tease out the energy of these archetypes in your life. Being in nature, allowing their messages to filter through the leaves is another way.

The framework I've shared with you in this book has been something I have spent over a decade refining. It will support your first steps, and you will find your own way from there. Enjoy where it leads you!

Kim Gould is a leading innovator and practitioner of Human Design. Working with multiple layer charts, Holographic HD life cycles, asteroids, and feminine DNA, Kim brings a more fluid approach to Human Design.

She loves to write about the emergent edge of current events and consciousness through a Human Design lens, as well as teaching, mentoring, coaching, and professional training.

Find out more about Kim at LoveYourDesign.com.

For more great books from Human Design Press
Visit HumanDesignStore.com

If you enjoyed reading *Asteroids in Human Design: Awakening the Feminine Archetype,* and purchased it through an online retailer, please return to the site and write a review to help others find the book.

Printed in Great Britain
by Amazon

11075780R00088